RICH STROUSS

FUNDAMENTALS
OF
METAL CASTING

To Edwina

This book is in the

ADDISON-WESLEY SERIES IN METALLURGY AND MATERIALS

———————

MORRIS COHEN

Consulting Editor

FUNDAMENTALS
OF
METAL CASTING

by

RICHARD A. FLINN

Professor of Metallurgical Engineering

Cast Metals Laboratory

University of Michigan

ADDISON-WESLEY PUBLISHING COMPANY, INC.

READING, MASSACHUSETTS · PALO ALTO · LONDON

PREFACE

The cast metals (foundry) industry has advanced further in the past ten years than in its entire history. This progress has been due to the concerted application of the sciences of physics, chemistry, metallurgy, and ceramics to foundry problems. Some striking examples are the development of ductile cast iron matching the strength of steel, of heat-resistant cast steels for severe jet-engine service, of new casting methods and mold materials giving accuracies within a few thousandths of an inch, and of refining techniques which permit the use of the lowest-grade scrap in the production of high-quality refined metal.

The purpose of this text is not merely to explain many of these new developments but, more importantly, to review the fundamentals of physics, chemistry, metallurgy, and ceramics which are of major importance in the foundry and then show how these are applied.

While modern physics and chemistry are very closely allied in foundry technology, we still think of physics traditionally as dealing with such topics as hydraulics, stresses, and heat transfer, while chemistry deals with the problems of analysis control, gas-slag-metal reactions, and the general field of chemical thermodynamics. These classical divisions provide a convenient method of arranging the material of this text.

Part I, which encompasses the physical aspects of casting liquid metal, begins with a discussion of the solidification of metals and its relation to temperature and composition. Next, the design of the mold is considered: first from the aspect of risering to provide an ample supply of liquid metal to compensate for liquid-to-solid shrinkage in the casting, and secondly from the standpoint of gating to deliver the metal most advantageously to the mold cavity and the risers. Then the stresses which develop as a result of thermal gradients during casting are analyzed in terms of the elastic and plastic strains set up during cooling. This topic is followed by a discussion of the mechanics of mold preparation and the physical properties of mold materials.

Part II, which treats the chemistry of liquid metals, starts with a discussion of simple gases in pure metals and then progresses to the more complicated cases in ferrous and nonferrous alloys, in which the separation of two constituents during solidification can lead to gas porosity or undesirable orientation of inclusions. The control of important elements in the standard chemical analysis of ferrous metals is described and forms the groundwork for the selection and control of melting furnaces. Chapter 13 contains a review of the principal calculations employed in Part II of the text. This procedure has the advantage of consolidating the much-used tables of heat content and entropy in one place and avoids long digressions in the other chapters to explain the computations.

The text has been written to serve two groups of readers: (1) the engineering students in metallurgy, and mechanical and industrial engineering, (2) the practicing foundry engineer.

The importance of a course of this type to the engineering student may be mentioned briefly here. Every professional engineer will readily admit that the best way to appreciate and retain a knowledge of the fundamental sciences is to apply it to actual problems. The ideal place for this course in the curriculum is the junior year, after the student has completed second-year college physics and chemistry, and preferably after a semester of physical chemistry. At this time it is important to provide an opportunity for laboratory work and application of the knowledge gained to such problems as solidification and other phase behavior, temperature measurement and calculation of thermal gradients, observation of mold-metal interactions, and fluid flow. Also, many experiments in chemical metallurgy, for example on gases in metals, slag-metal equilibria, and so forth, help fix the fundamentals in the student's mind. At Michigan, these are conducted on an informal basis, i.e., there are no recipes to be followed, and the only limitations are the ingenuity of the student and the proper safety precautions. Organized in this manner, the course provides a good basic starting point in physical, chemical, and mechanical metallurgy, as well as an important professional background.

The text is also written for the practicing foundry engineer who wishes to reach an overall understanding of fundamentals. In the foundry, as in other fields, there is a tendency for specialization at the risk of missing developments in other fields which can be profitably applied to the field of specialization. For example, many of the concepts of gating design that were developed for aluminum alloys are of equal importance for certain steels, cast irons, and copper-base alloys which are easily oxidized. Many of the principles of gas elimination are important to the casting of alloys in general.

In the past, many engineers have entered the foundry without realizing how the methods of their early training in science should be applied to their everyday problems. Many millions of dollars have been wasted on trial-and-error methods of cupola-, electric-, and other furnace operations and in the selection of mold materials and mold design. It is our hope that this review of the fundamentals of casting will help some of our hard-working colleagues in industry to refurbish their fine early education for a fresh approach to their daily problems.

There will probably be sincere criticism of the attempted wide scope of this text. But if a text is to be written about cast metals, it must be with an effort to embrace the most recent developments in all the sciences.

We have tried to avoid the qualitative approach, the primary cause of failure in many efforts at wide coverage. At the risk of omitting much of the usual material, the best *quantitative* treatments in our experience have been employed.

One final comment. The practical critic will miss much of the descriptive detail he may consider important. The author has had to make a deliberate allotment of space between telling "how" to make castings and "why" certain phenomena occur. It has seemed more important to emphasize the "why" because before this book is published, new core washes and new sand mixes will have outmoded many that are mentioned here. The "why," however, remains a permanent part of the engineer's background and enables him to understand or, even better, to develop the new materials and methods of the future.

I would like to acknowledge the many contributions to the text: the sympathetic and helpful criticism of Professor Morris Cohen, the help of my colleagues, Professors G. A. Colligan, R. W. Kraft, W. B. Pierce, W. A. Spindler, D. A. Sponseller, and especially Professor Paul K. Trojan, who reviewed the text, and the skillful editorial and art work of the Addison-Wesley staff.

I would also like to acknowledge the excellent first-hand industrial experience I obtained, first at the International Nickel Company with Mr. D. J. Reese and later at the American Brake Shoe Company, which provided much of the background necessary for the writing of this text. The material is published here for the first time.

Finally I appreciate the permission to publish three figures from Pfann, *Zone Melting* (John Wiley and Sons), and a number of figures from *Transactions of the American Institute of Mining and Metallurgical Engineers.*

Ann Arbor   R. A. F.
December 1962

CONTENTS

PART I

THE PHYSICAL ASPECTS OF CASTING:
MOLD DESIGN, PRODUCTION, AND MATERIALS

PART II

THE CHEMISTRY OF LIQUID METAL;
CONTROL OF COMPOSITION, MELTING, AND REFINING

PART I

The Physical Aspects of Casting: Mold Design, Production, and Materials

CHAPTER 1

INTRODUCTION

Before plunging into the uncharted land before him, the experienced explorer will look for a nearby mountaintop and spend considerable time studying the path ahead. In the same way, a seasoned engineer will take what may seem to a novice an inordinate amount of time to study a problem as a whole from a distance before becoming immersed in the details of its solution. Similarly, in this chapter we shall examine castings and the cast-metals industry from a general point of view before taking up the details. For the engineering student this approach is quite important, because the subject is rather different from any he has encountered. In most courses a particular subject is isolated and so simplified. For example, we do not often make intensive use of mathematics, chemistry, and physics in the same course, but to acquire a real understanding of a successful casting, or better yet of an unsuccessful one, requires the concurrent application of all the basic sciences.

In this text we shall break down the problem of the proper engineering of a casting into several logical portions. However, it must be appreciated that this approach is, in a way, artificial, and that a successful casting is attained only by concurrent success in all the branches of its manufacture. For example, it is of little solace to an anxious consumer of aircraft castings to learn that a foundry has been able to produce an urgently needed casting with exactly the desired metal structure but not within proper dimensional tolerances!

To appreciate the breadth of the problem, let us begin with a general view of the purpose and manufacture of castings and then proceed to a study of the cast-metals (foundry) industry as a whole. With this background, the need for the detailed discussions of later chapters and their interrelation will become apparent. To make one last reference to the analogy of the explorer: just as remembering the general view from the mountaintop helps the explorer to understand the need for cutting through a dense forest, so we hope that the content of this chapter may motivate the student to review the basic sciences when some of the more difficult material of later chapters is encountered.

1–1 What is a metal casting? A metal casting is a shape obtained by pouring liquid metal into a mold or cavity and allowing it to freeze and thus to take the form of the mold. This is the fastest and often the most economical method for obtaining a part of any desired composition. It

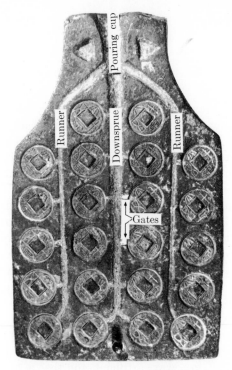

FIG. 1–1. Chinese money mold, about 2000 B.C. (approx. ⅓ size).

FIG. 1–2. Missile rotor (× 1⅓).

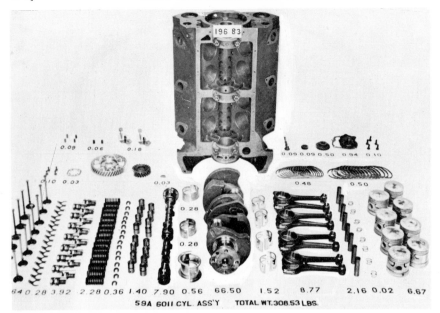

FIG. 1–3. Over ninety percent of these automotive V-8 engine parts are cast.

is a fascinating sight in any foundry to see a pile of ingot or scrap metal converted to an array of working metal parts, often in a matter of hours. The Chinese money mold of Fig. 1–1 is an interesting example of this method, dating from 2000 B.C. When this mold is assembled, book fashion, with a mating section, liquid brass poured into the *pouring cup** at the top passes into the *downsprue* through the *runner* and enters the casting cavities through *gates*. This particular mold is called a *permanent* mold because it is re-usable; many molds are made of bonded ceramic materials, such as sand, and are used only once. As a matter of fact, this permanent mold was made in a bonded sand mold, and so is itself a casting.

Some modern castings that illustrate the versatility of foundry techniques can be seen in Figs. 1–2 through 1–4. The rotor from a missile shown in Fig. 1–2 is made of a heat-resistant alloy, to withstand the hot gases which impinge upon the blades. The dimensions of the blade are cast to within 0.003 in., since the material is difficult to machine and the shape is complex. A large percentage of the automobile parts shown in Fig. 1–3 are also cast.

Castings are often used as components in a complex structure. In the early stages of the development of the British Sapphire jet engine, the assembly shown in Fig. 1–4 was machined from large forgings; approxi-

---

* Complete definitions are given in the Appendix.

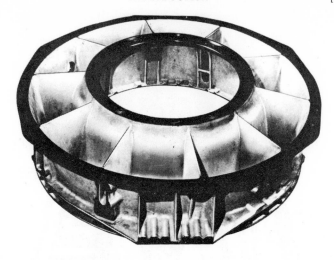

FIG. 1–4. Center main bearing support in a Sapphire jet engine (size of bearing in photograph is one-fifteenth of original size).

mately 6000 man-hours were required to produce this structure. Later, the process was changed to cast the two annular sections separately and precisely and then weld stainless-steel struts in position; the part was then produced in a tenth of the time and at greatly reduced expense. The performance was also improved by extending the range of operating temperatures through the use of an alloy which could be readily cast (austenitic ductile iron).

With these few examples as a background, let us proceed now to a view of the industry as a whole.

1–2 A survey of the foundry industry. (a) *Production.* Because of the basic simplicity of the casting process, almost any metal or metal alloy can be cast. The figures in Table 1–1 are an indication of the annual production of some of the common alloys. In addition to the production of great tonnages of castings of the more common metals, castings of the newer, highly reactive alloys such as titanium are being made. A new family of alloys, called *ductile iron,* is rapidly increasing in importance. The figure for its annual production is not yet available, however.

In assessing tonnage, we should realize that the total in Table 1–1 represents finished or nearly finished products and is equivalent to about twice the value in ingot tons. The cast-metals industry has been ranked fifth in the capital-goods group.

(b) *Products and producers.* The castings just discussed illustrate only a very small cross section of the variety of casting uses. Every important

TABLE 1–1

Alloy	Shipments of castings in 1960, millions of net tons
Gray cast iron	12.7
Cast steel	1.8
Malleable iron	0.8
Brass and bronze	0.5
Aluminum alloys	0.38
Zinc-base alloys	0.38
Magnesium alloys	0.03
	Total 16.6

capital-goods industry uses castings either directly or indirectly in production and processing. There are over 6300 foundries in the United States and Canada. The state of Ohio leads with 577 foundries, followed by the states of California, Pennsylvania, Illinois, Michigan, Indiana, and Wisconsin.

1–3 The casting problem defined. The preceding survey of castings and of the foundry industry indicates a wide variety of both products and producers. How is it possible then to cover adequately in one text the fundamentals of casting? It is unfortunate that many foundry personnel are unaware of the common fundamental problems of the industry and so are able to communicate with other specialists only in such specialized areas as "green-sand squeezer castings" or "zinc die castings." This isolation of development work, coupled with the great number of producers, has been a source of weakness which has retarded the application of many modern developments in related sciences to problems of the foundry industry. Therefore our approach here will be to avoid isolated discussions of such things as "gray-iron practice" or "light-metal castings," and instead to gather the best basic developments in each field for comparison and interpretation. For example, although important work in the field of gating has been done with aluminum alloys, more complete work on risering and solidification has been accomplished with the ferrous metals. The principles underlying these and other basic developments are broadly applicable and should be widely used.

To illustrate the general nature of the casting problem, to introduce some nomenclature that may be new to the reader, and to outline the path which will be followed in subsequent chapters, let us consider an actual casting familiar to us all, a railroad-car wheel (Fig. 1–5). The same

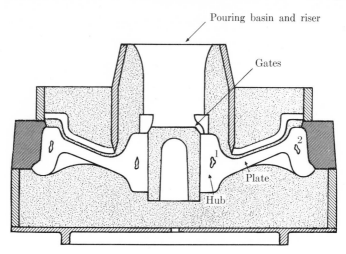

FIG. 1–5. Railroad-car wheel mold; green-sand practice.

general principles that we shall discuss here will apply equally well to a two-ounce gas turbine blade or to a 15-ton tank hull.

This casting problem, as any other, may be divided into two general parts: first, the design and production of the mold, and secondly, the melting, refining, and pouring of the liquid metal. In studying this case we shall point out typical questions that arise in most casting problems. The order of presentation is merely for clarity; in actual practice all these factors should be kept in mind at once.

1–4 The design and production of the mold. (a) *The mechanism and rate of metal solidification.* For the risers to be placed and proportioned properly, the crystallization pattern and shrinkage in volume accompanying solidification must be known. These factors vary considerably, depending upon the chemical composition of the metal and the thermal gradients in the mold.

(b) *Heat transfer during solidification* (*risering*). After the solidification process is understood, one can proceed to study the control of shrinkage porosity by application of heat-transfer principles. In the example of the wheel, the volume of the liquid metal is greater than that of the solid metal, and therefore voids will occur at regions 1 and 2 (Fig. 1–5) unless steps are taken to change the natural thermal gradients. We need to know what can be done to change the heat-transfer conditions by the use of liquid-metal reservoirs (risers) and of other mold materials such as metal chills.

(c) *The flow of liquid metal.* The problems involved here are illustrated by the following questions: What are the calculations leading to the selection of the proper pouring temperature? On the one hand, we must avoid destruction of the mold by liquid metal at excessively high temperatures, and on the other, we must prevent solidification of the metal stream before the mold is completely filled. How shall the channels for delivering the metal (the *gating* system) be designed? In one method, as illustrated in Fig. 1–5, the metal is delivered to the *pouring basin* and passes through small, narrow channels, or *gates,* into the mold cavity. While this system is satisfactory for steel, the turbulence would be intolerable for an easily oxidized metal such as aluminum, since the casting would entrap accumulations of aluminum oxide or *dross.* For aluminum it is necessary to design a gating system to introduce the metal in a quiet stream, preferably from the bottom of the mold. However, this method is not restricted to aluminum; in a very successful, recently introduced process for making a steel wheel, the metal is introduced from the bottom of the mold by a special pressure-pouring device [1].*

(d) *Stresses in metal in the solidus temperature range.* Even though a sound casting has been obtained, the thermal gradients developed during solidification can cause serious stresses that lead to hot tears at the temperature of solidification. Can these be reduced by changes in casting shape or mold construction?

(e) *Stresses in the elastic range.* The thermal gradients in the casting at high temperatures can also lead to high stresses in the cold casting. In the wheel shown, the slower cooling rate of the hub may lead to high radial tension in the plate or web of the final casting. The methods for relief of this stress, or for the development of beneficial residual stresses, are of basic importance.

(f) *Mold materials and production methods.* Here it is necessary to select the most economical materials for mold construction. Which molding method will provide satisfactory cavities at lowest cost?

1–5 Melting, refining, and pouring of liquid metal. The proper selection of the methods for melting and refining metal of the desired composition, another significant facet of the general casting problem, requires careful study of the following areas.

(a) *Gases in metals.* During melting, porosity in castings is often produced by solution of gases in liquid metal which are less soluble in the solid metal and therefore precipitate as bubbles, leading to holes. On the other hand, certain dissolved gases, such as nitrogen, can be helpful in producing special metal structures. How can the broad background of physical chemistry be applied to the problem of selecting the proper melting condition for a given metal?

---

* Numbers in brackets are keyed to the references at the ends of the chapters.

(b) *Control of common elements.* When service tests have shown a metal of a certain composition to be most desirable, we need to determine what combination of scrap, pig, flux, furnace conditions, and temperature is necessary for reliable, economical production.

(c) *Selection and control of melting furnaces.* Given the choice of several slag-metal reactions, we need to decide which furnace or combination of furnaces should be selected to provide the desired rate of delivery, temperature, batch size, and composition of metal.

This preliminary illustration may seem overly complex to the beginner and oversimplified to the foundry engineer. It should, however, establish the point that the development of any casting for engineering use depends upon the successful application of physics, chemistry, ceramics, and metallurgy.

In the chapters to follow, we shall focus our attention upon the individual facets of the casting problem as just outlined.

References

1. H. H. Hursen, "Pressure Pouring and Graphite Permanent Molds Used in the Production of Steel Car Wheels," *Trans. A.F.S.* **63**, 367–372 (1955).

2. *Transactions of the American Foundrymen's Society (A.F.S.).* Contains many basic papers on solidification, thermal gradients, heat transfer as well as practical papers dealing with sand control, foundry layout, etc. Papers are published first in *Modern Castings*, the monthly magazine of the A.F.S.

3. *The Foundry* contains many very readable articles on foundry fundamentals and practice as well as important industrial statistics.

4. *Iron Age, Steel,* and *American Metal Market* give data on material costs and processes related to the foundry industry.

Problems

1. Look up an article on the production of a particular casting in reference 2 or 3. Make a sketch of the casting, labeling risers, gating system, downsprue, cope side, drag side, runner, vents.

2. Prepare a short report on the above casting, organizing the data and methods according to the breakdown given in Sections 1–4 and 1–5.

CHAPTER 2

MECHANISM AND RATE OF SOLIDIFICATION OF METALS AND ALLOYS

2–1 General remarks. The few seconds or minutes during which a casting solidifies occupy but a relatively brief period in its entire production history. During this short time, however, the original crystal structure of the casting is formed, the backbone upon which many properties depend. Also, in this interval major flaws such as shrinkage porosity, hot tears, and seams can be prevented, depending upon the care with which the solidification has been planned. Hence it behooves us to spend a substantial amount of time reviewing in general terms what is known about the crystallization of metals and alloys from the liquid state. We can then discuss the crystallization patterns in actual castings and, following these discussions of the *mechanism* of solidification, we can take up the factors affecting the *rate* of solidification and the heat-transfer calculations involved. All this material will be of immediate use in the next three chapters, dealing with riser calculations (Chapter 3), gating (Chapter 4), and fluidity (Chapter 5).

2–2 Solidification of pure metals. Why does a liquid metal solidify? Basically the reason is that the arrangement of the atoms in a solid crystal is at a lower free energy* than that of the same atoms in a liquid state. Above the freezing point, however, the liquid state is more stable. *At* the freezing point there is no driving force in either direction; in other words, the change in free energy is zero, and we have equilibrium.

From this brief analysis we see that a crucible of liquid metal will not quite begin to freeze at the freezing point (since there is then no driving force to produce solidification), but will begin to freeze at some lower temperature. The further the metal is cooled below the liquid-solid equilibrium temperature, the greater is the driving force to solidify.

The reason that solidification does not begin immediately when metal is supercooled† a fraction of a degree below the liquid-solid equilibrium temperature is that energy is required to produce the new surfaces of the crystals being formed, just as it takes energy to blow and enlarge a soap bubble against the force of surface tension.

---

* Free energy is discussed in more detail in Part II of the text.

† The terms *undercooling* and *supercooling* are used interchangeably; both signify the cooling of a phase to below the equilibrium transformation temperature, for example, liquid water at 30°F (−1°C).

11

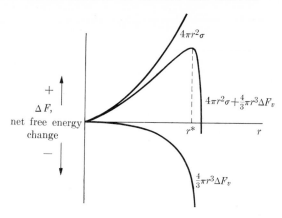

FIGURE 2-1

Because of this energy requirement, it is sometimes possible to cool and maintain a liquid far below its freezing temperature under carefully controlled conditions. Let us now express quantitatively the factors in the balance between these forces affecting crystal *nucleation*.

There are two types of nucleation: *homogeneous* and *heterogeneous*. We may define homogeneous nucleation as the formation of a new phase without the help of special nucleation sites. By heterogeneous nucleation in the solidification process, we mean that the solid phase crystallizes on foreign nuclei. Although most solidification or crystallization is initiated by foreign nuclei, it is helpful to analyze the homogeneous case first and then modify the analysis for the heterogeneous type.

(a) *Homogeneous nucleation.* The temperature at which homogeneous nucleation occurs is always below the equilibrium freezing point because it is necessary to overcome the surface-tension forces which impede nucleus growth. The energy tending to produce a nucleus of solid is the difference in free energy per unit of volume between the liquid and solid phases. This quantity, called the bulk free-energy change, is designated as ΔF_V, and is usually expressed in calories/cm³ for a given temperature of supercooling. At all temperatures below the freezing point, ΔF_V is negative; i.e., the solid phase is then more stable, and therefore has a lower free energy than the liquid phase. The bulk free-energy change associated with the formation of a spherical particle of radius r is $\frac{4}{3}\pi r^3 \Delta F_V$. The energy required to create the new surface is a function of the surface tension, σ, in ergs/cm², or $4\pi r^2 \sigma$. The net free-energy charge, ΔF, is then

$$\Delta F = 4\pi r^2 \sigma + \tfrac{4}{3}\pi r^3 \Delta F_V. \tag{2-1}$$

The surface-energy term is positive in contrast to the bulk free-energy term, resulting in the curve of ΔF versus particle size shown in Fig. 2-1. Since different powers of r are involved, the curve will show a maximum at a critical radius r^*. Note that once this radius is exceeded, further growth results in a decrease of free energy and can proceed spontaneously at the temperature of supercooling. If a particle has a radius smaller than r^*, it will redissolve because this decrease in size reduces the free energy. The size of the critical radius may be found by differentiating Eq. (2–1) with respect to r and setting the result equal to zero:

$$r^* = - \frac{2\sigma}{\Delta F_V},$$ (2–2)

where r^* defines the critical nucleus size.

The significance of this relationship is that for the critical radius to become very small, ΔF_V must become large in a negative sense; in other words, severe supercooling is needed for homogeneous nucleation. (The term σ does not change greatly with temperature.) However, ΔF_V increases with supercooling and therefore causes a very small nucleus to be stable.

Obviously, a nucleus formed spontaneously in the melt by random atomic motion will probably only be of the size of a cluster of a few atoms; therefore extreme supercooling will be necessary for it to exceed the critical nucleus size and grow spontaneously. Walker [12] has shown that melts of several hundred grams of nickel can be supercooled to 500°F (-296°C) below the freezing point by taking precautions to avoid foreign nuclei.

(b) *Heterogeneous nucleation.* Most actual castings crystallize by heterogeneous nucleation, the basic reason being that if the new phase can find a foreign particle to grow upon, it can in effect adopt the relatively large radius of the particle as its own. This means that only a slight degree of supercooling is needed in comparison with that needed for homogeneous nucleation. Quantitatively, the relation depends upon the degree to which the new phase "wets" the foreign particle. If there is no attraction between the atoms of the foreign particle and those of the precipitating phase, then nucleation is not helped. The wall of a mold usually provides many heterogeneous nucleation sites. The best nucleus, of course, is a particle of the precipitate itself. For example, it is possible to grow large single crystals of metals by introducing a small crystal of the metal itself into a melt as it cools through the freezing temperature.

To summarize the conditions prevailing during the solidification of a *pure metal*, let us now review what occurs when a metal of freezing temperature T_E freezes on being poured into a mold (Fig. 2–2). If it is poured rapidly, there is no appreciable thermal gradient† at time t_0. Then, due to the heat transfer at the interface, a thermal gradient develops in the metal. The interface temperature falls slightly below T_E, and at time t_1

---

† Thermal gradient is defined to be the change in temperature as a function of distance.

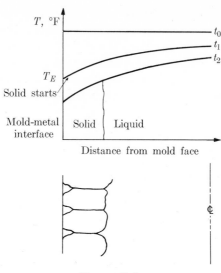

FIGURE 2–2

crystallization begins at the interface (heterogeneous nucleation). (It must be recalled here that there is a direction of preferred growth in a crystal, as in the case of the growth of ice crystals on a window pane.) Those nuclei which happen to be formed with the preferred growth direction perpendicular to the interface grow more rapidly than crystals of other orientations. This leads to the growth of columnar grains advancing toward the center of the ingot. The final structure therefore consists of a thin layer of randomly oriented grains at the mold surface and columnar grains extending to the centerline. In this situation, there are no randomly oriented grains at the center of the ingot.

2–3 Nucleation and growth in alloys. The solidification of alloys differs in three principal ways from that of pure metals: (1) usually, the freezing of alloys occurs over a temperature *range;* (2) the composition of the solid which separates first is different from that of the liquid; (3) there may be more than one solid phase crystallizing from the liquid.

(a) *Solid-solution alloys.* If we dissolve metal B in liquid metal A to form a liquid alloy, we can describe the crystallization by means of a phase diagram, as shown in Fig. 2–3. In most cases the freezing point of the metal A is depressed as B is added, as indicated by the liquidus line. In addition, the alloy, of composition C_0, for example, does not freeze at a single temperature, but instead over a temperature range. (The temperature at which freezing begins is called the *liquidus,* and the temperature at which freezing is complete is called the *solidus.*)

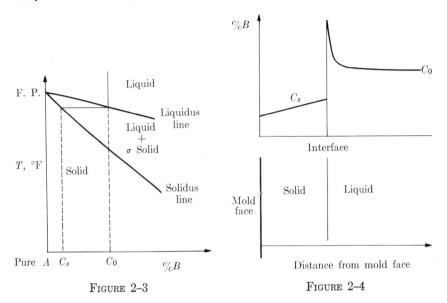

FIGURE 2–3   FIGURE 2–4

Another important fact is that as the metal solidifies, the composition of the solid is not the same as that of the parent liquid, but is richer in the metal A. The B-atoms in the crystal structure of A are in *solid solution,* and are distributed among the A-atoms in an atomic dispersion. We shall discuss now how these effects lead to a solidification pattern different from those described for the pure metal.

If we begin with a homogeneous melt of composition C_0, the first crystals to precipitate are of composition C_s. If solidification is fairly rapid, so that *no diffusion occurs,* the liquid at the interface becomes richer in the solute (B) than the liquid away from the interface. The variation of $\%B$ with distance is shown in Fig. 2–4 for temperatures between the liquidus and solidus.

Assume that the mold is filled rapidly, as in the previous example illustrating the case of a pure metal so that there is no temperature gradient at the start when $t = 0$, as shown in Fig. 2–5.

Consider the liquidus temperature of composition C_0 as T_E. As previously, we establish a thermal gradient and supercool somewhat, before crystallization begins at time t_1. Then consider the situation at t_2 (after appreciable solid has separated). A temperature gradient similar to that of the pure metal is assumed. The difference between this situation and that of the pure metal is that the liquidus temperature of the remaining liquid now varies with distance from the interface, as shown by the dashed line. Near the interface, due to build-up of component B in the liquid, the freezing temperature is considerably lower than T_E. Well

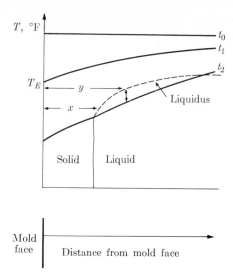

T, °F

t_0
t_1
t_2

T_E

y

x

Liquidus

Solid   Liquid

Mold
face   Distance from mold face

FIG. 2–5. Solid lines show actual temperatures in the mold at t_0, t_1, and t_2.

away from the interface, the equilibrium freezing temperature is still T_E, if we assume no diffusion.

Note now that a region of liquid at point x has a higher freezing point (dashed line in Fig. 2–5) than the liquid *at* the interface, while the actual metal temperature at x is also higher than at the interface. However, the difference between the *liquidus* and *actual* temperatures at x is greater than the corresponding difference at the interface. The liquid at x is then said to be *constitutionally* supercooled.

To emphasize this point, let us consider two separate crucibles of metal, one of the liquid composition at the interface, say 10% *B*, and the other of the composition at point x, say 5% *B*. Let the liquidus of the interface liquid be 1000°F (538°C) and that of the other 1050°F (566°C). Now suppose that the actual thermal gradient in the metal results in an interface temperature of 990°F (532°C) and a temperature at point x of 1000°F. It is clear that the liquid at the interface is only 10°F (6°C) below its liquidus, while that at point x is 50°F (28°C) below. The bulk free energy of the latter, which is available for crystallization, is therefore greater.

The effects of constitutional supercooling upon the crystallization are of three types, depending upon the degree of supercooling.

(1) If there is only minor supercooling, certain preferred regions of the interface will protrude as spikes into the supercooled region and, once started, will grow more rapidly than neighboring regions (Fig. 2–6). This will happen both because the driving force for freezing is greater in

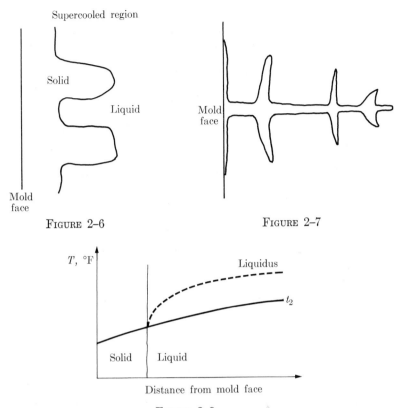

FIGURE 2–6

FIGURE 2–7

FIGURE 2–8

the supercooled region and because these spikes will reject solute at their sides, thus delaying freezing of the side regions. The spikes may result in the formation of a honeycomb structure.

(2) If supercooling is greater, the spikes tend to form side arms, producing a *dendritic* structure (Fig. 2–7).

(3) Finally, in the case of extreme supercooling, the temperature difference, $T_{\text{liquidus}} - T_{\text{actual}}$, which is a maximum at some distance y in Fig. 2–5, may become large enough to lead to independent crystallization. In this way, randomly oriented (equiaxed) grains may be encountered toward the central part of an alloy ingot. An extreme situation of this type is represented by Fig. 2–8. It should be mentioned that the relative solubility of the solute is shown by the factor k, where

$$k = \frac{C_s}{C_0}.$$

Typical values of k lie in the range 0.01 to 0.5. When both k and the

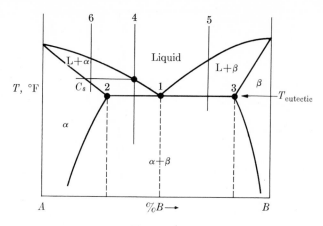

FIGURE 2-9

thermal gradient are low, even the metal at the center of the casting can be constitutionally supercooled. On the other hand, it should be noted that the greater the thermal gradient, the greater the possibility for columnar growth. This tends to prevent dendritic growth and random crystallization ahead of the freezing interface and thereby facilitates metal flow for feeding of solidification shrinkage. In other words, the greater the thermal gradient, as in chilled castings, the simpler the feeding problem.

There is one other effect, called "mass feeding," which should be mentioned for the sake of completeness. When equiaxed crystals are nucleated randomly ahead of the solidifying interface, the tendency of these crystals is to sink in the liquid because of the greater density of the solid. This effect results in less porosity at the bottom portion of a section than in the upper part, and sometimes also leads to layering of the shrinkage voids.

(b) *Eutectic alloys.* In this case, under equilibrium conditions, two different solids are formed simultaneously from the liquid, and the reaction is completed at constant temperature. In Fig. 2–9 liquid metal of composition 1 reacts to form solids of compositions 2 and 3, respectively, at the eutectic temperature.

If the composition of the liquid is *hypoeutectic,* i.e., lies to the left of composition 1 but contains more B than point 2, the crystallization occurs first as discussed in (a) above for single solid precipitation, and then eutectic formation takes place. In other words, liquid of composition 4, upon cooling below the liquidus temperature, precipitates the α-phase of composition C_s. As this single-phase precipitation continues, the melt becomes richer in element B and finally attains the eutectic composition at

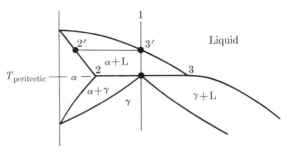

FIGURE 2-10

point 1. Then both solid phases, α and β, precipitate simultaneously. In analogous fashion, if the melt were of composition 5 (called hypereutectic), single-phase precipitation of the B-rich β-phase would occur first, followed ultimately by the eutectic formation.

The composition limits within which the eutectic reaction occurs are widened by nonequilibrium freezing conditions. For example, the eutectic reaction may be encountered in an alloy of composition 6 if it is cooled rapidly. Diffusion of B is not rapid enough to allow the solid first formed to increase in % B as called for by the solidus line. The liquid becomes correspondingly richer in B and eventually attains the eutectic composition.

(c) *Peritectic reaction.* The peritectic reaction also involves relationships among three phases at constant temperature, but here a solid and a liquid react on cooling to form a new solid, as shown in Fig. 2-10. If the liquid of composition 1 is cooled, the solid of composition 2′ first precipitates. Under equilibrium conditions the composition of all the solid follows the solidus line from 2′ to 2, and the liquid takes the liquidus path from 3′ to 3.

At the peritectic temperature, the solid, α, of composition 2 and the liquid of composition 3 react to form a third phase, solid γ of composition 1. If the composition of the original melt is to the left of composition 1, there will be an excess of α after the peritectic reaction. It should be pointed out that the peritectic reaction is very sluggish compared with precipitation from a liquid. The explanation is that a *solid* phase α reacts with liquid to form a new phase γ. The new phase will coat the remaining α. For the reaction to continue, diffusion through the *solid* γ is required, a relatively slow process.

2-4 The solidification of actual castings. Let us pass on now to observations of the solidification of actual castings. As an example we shall use the iron-carbon system, which includes some of our most important

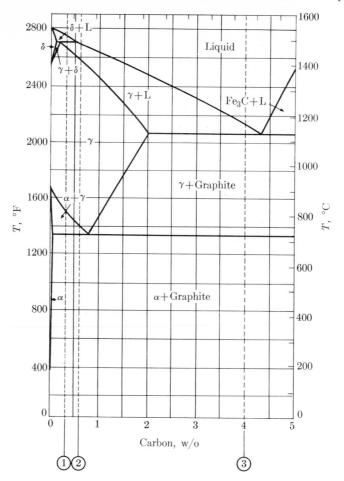

Fig. 2–11. Iron-graphite diagram.

alloys, such as steel, malleable iron, gray iron, and ductile iron. By selecting various carbon contents we can illustrate all of the types of freezing we have just discussed.

At this point we shall digress to answer the question, "How do we know when crystallization is occurring inside a mold?" A variety of techniques have been used, but thermal analysis is the most common. If we embed a thermocouple in a fused silica protection tube, a graph of temperature versus time, a "cooling curve," will yield considerable information concerning the phase changes taking place in the metal. Let us review the cooling curves that we would expect from iron-carbon alloys of three types: (1) pure iron (0% C), (2) solid solution (0.6% C),

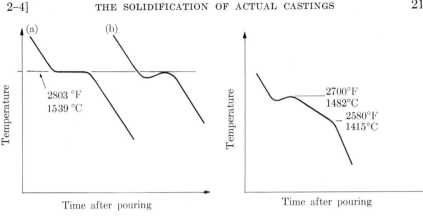

FIGURE 2–12   FIGURE 2–13

and (3) eutectic (4.3% C), as shown in the phase diagram of Fig. 2–11. In a melt of pure liquid iron, the temperature falls continuously until crystallization begins at about 2800°F (1539°C), Fig. 2–12(a). At this point, the heat of solidification is evolved, and the temperature remains constant until solidification is complete. Depending upon nucleation conditions, the graph may show thermal supercooling effects, Fig. 2–12(b). In the case of the solid solution, the first crystals (austenite) start to grow from the liquid at about 2700°F (1482°C), continuing to 2580°F (1415°C), when the casting is completely solid. The heat of solidification is evolved continuously during this period, and therefore the cooling rate is retarded until solidification is complete. Some supercooling usually ensues before initial crystallization, leading to a horizontal portion of the graph or even a temperature rise (Fig. 2–13). The arrest in the cooling curve is due to the fact that the melt is considerably below the liquidus at the time at which solidification begins. Once solid nuclei develop, an appreciable quantity of solid precipitates. In a melt of iron containing

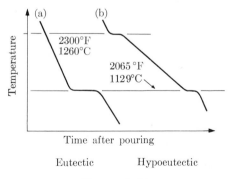

FIGURE 2–14

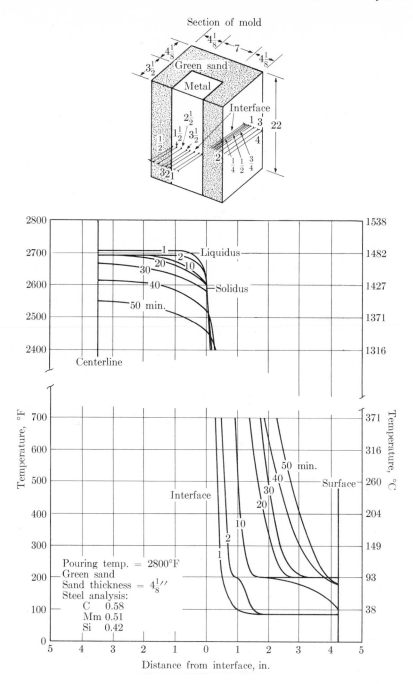

FIG. 2–15. Temperature-distribution curves for 0.6% carbon steel (alloy 2).

4.3% carbon, the liquid cools to about 2065°F (1129°C) before solidification begins. At this point, with slow cooling, both austenite (2.0% C) and graphite crystallize at constant temperature until the casting is solid. This reaction results in a curve similar to that of a pure metal, as in Fig. 2–14(a). If, however, the composition contains 3% carbon, for example, the curve will show a change in slope from the liquidus to the eutectic temperature and then a long arrest, as can be seen in curve (b) of Fig. 2–14.

Now let us return to our observations of actual castings, selecting a simple design of square cross section 7 inches × 7 inches and 21 inches long (Fig. 2–15). A large riser, not shown in the figure, is placed at the top of the casting to eliminate shrinkage effects. Thermocouples are located at intervals in the bar and in the green-sand mold on a central plane to eliminate end-cooling effects. Cooling curves are then obtained at each station. From these curves the temperature distribution throughout the mold can be determined at any given time, as illustrated in Fig. 2–15, which shows the temperature-distribution curves for a 0.6% carbon steel, for various times after pouring. Note that these graphs are not cooling curves but are derived from them.

Let us consider the thermal gradient at one minute after pouring. Note first that the stations within $\frac{1}{2}$ inch of the mold surface are below the liquidus, while those deeper within the bar are above. This indicates that a freezing front develops at the surface and moves inward, as would be logically expected. The heat being transferred through the mold-metal interface raises the temperature of the sand mold.

At the centerline of the bar, the liquidus temperature is reached about two minutes after pouring and remains constant for 22 minutes. Conversely, during the later periods the temperature falls rather steadily. The isothermal arrest is caused by the liberation of the heat of solidification from the outer portions of the bar, which produces a thermal barrier to heat transfer from the center. As the heat of solidification is transferred from the outer regions through the mold wall, the thermal block (knee of the curve) moves inward toward the centerline.

It is also interesting to note a similar thermal block in the green-sand mold. The zone between the 1-inch and $3\frac{1}{2}$-inch stations remains at 212°F (100°C) for a long period because of vaporization of the water in the sand and the corresponding latent-heat requirement at constant temperature. This phenomenon is analogous to the arrest caused by the heat of solidification in the metal.

We now focus our attention on the end of the freezing period. The solidifying metal at the mold surface falls below the solidus temperature (becomes *completely* solid) about 30 minutes after pouring. However, the centerline temperature has been below the *liquidus* since 24 minutes

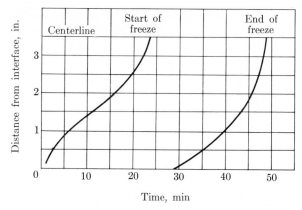

FIG. 2–16. Progress of freezing in steel casting of Fig. 2–15 [1].

after pouring, which indicates that some crystallization has been occurring at the center of the casting for six minutes prior to the complete
solidification of the surface. In other words, a mixture of liquid + solid
has extended from the surface to the center for six minutes.

These facts are summarized in the graph of Fig. 2–16. The most important point to note [1] is that the steel does not freeze with a smoothly
advancing front, but rather a start-of-freeze wave passes from the surface to the center, followed much later (~ 20 minutes) by an end-of-
freeze wave. In the· interval between these waves, the bar contains a
mixture of solid crystals (usually dendrites) and liquid. The existence of
this region is an impediment to the delivery of liquid metal from the riser
reservoir, and, if the difficulty is not overcome, shrinkage voids will be
found in the casting. We shall treat this problem quantitatively later;
here we shall pass on to the observation of actual freezing curves in an
alloy containing a eutectic.

Cooling of an alloy containing a eutectic [2]. A cast iron of the following composition was chosen for study:

C	Mn	P	S	Si	Fe
3.04	0.53	0.01	0.05	1.39	balance

For this material we would expect a cooling curve somewhat similar to
that of case (b) in Fig. 2–14. However, the eutectic does not freeze at
constant temperature in commercial cast iron because the presence of
other components, especially silicon, leads to a eutectic freezing range
(about 2109 to 2089°F, or 1154 to 1143°C). The actual cooling curves
are shown in Fig. 2–17(a). The curves for the variation in temperature at
selected times after pouring are developed in the same manner as for the
steel we discussed earlier.

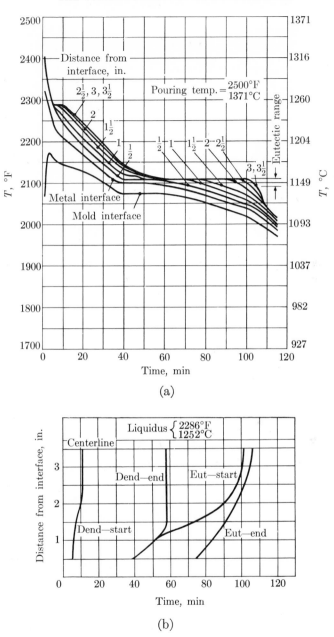

Fɪɢ. 2–17. Cooling curves showing solidification of gray iron [2]. Same test casting as shown in Fig. 2–15 [1].

Fig. 2–18. Riser of hypereutectic iron from which metal was drained prior to final solidification (approx. one-third size).

In this cast iron, two thermal arrests are evident. The first is caused by the heat of solidification of the austenite alone, as in the case of steel. The second is produced by the heat of solidification of the eutectic (austenite + graphite) over a narrow temperature range. From these effects, the beginning and the end of each reaction are plotted (Fig. 2–17b). The separation of the "dendrite end" curve from the "eutectic start" curve, which may be criticized, is caused by the arbitrary selection [2] of a temperature two degrees above the eutectic range as the "end" of austenite dendrite formation and the selection of a temperature two degrees below the eutectic as the "start" of the eutectic formation. Although only a four-degree spread is involved, a relatively long time is required for the inner portions of a casting to pass through this range, the delay being due to the thermal arrest which is caused by the heat of eutectic solidification evolved in the outer regions. In actual practice, the crystallization of the eutectic austenite can and does take place upon the already existing austenitic dendrites, and is therefore a continuous process. The arbitrary selection of temperature for "end of austenite dendrite" and "start of eutectic" formation does, however, serve to emphasize the effects of the two types of reactions upon the solidification phenomena (Fig. 2–17b).

2–5 The effects of mold material and alloy composition upon freezing pattern (feeding resistance). We have established that, at least for the $7 \times 7 \times 21$ inch bar just discussed, there is some solid material at the center of the casting for a considerable interval before final solidification.

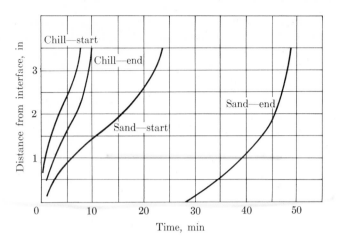

FIG. 2–19. Comparison of solidification rate of chilled and sand-cast 0.6% carbon steel poured at 2800°F (1538°C); mold as in Fig. 2–15 [1].

This observation is confirmed experimentally by the casting shown in Fig. 2–18, from which liquid metal was drained at a selected point in the crystallization process. It is obviously more difficult to deliver liquid metal through such a network of dendrites to the locations undergoing final solidification than if only liquid were present and solidification occurred in a single wave. Thus it is important to examine the effects of the mold material and of the metal composition upon the extent of this dendritic pattern.

We shall use two criteria to evaluate these effects: the relative positions of the start- and end-of-freeze curves and a recently introduced parameter [6], the "centerline feeding resistance."

(a) *Effect of mold material.* If a cast-iron ingot mold is substituted for the sand mold previously discussed, the higher conductivity of the mold material will induce more rapid freezing. The curves of Fig. 2–19 indicate this effect, but what is more important, they show that the freezing pattern is altered. In the sand casting, a mushy condition, i.e., liquid + solid, existed throughout for a six-minute period, namely from 24 to 30 minutes after pouring. Conversely, this condition did not occur in the chilled casting; only a narrow band of liquid + solid progressed through the liquid ahead of the completely solid material. For example, at the time at which freezing *began* at the centerline, two inches of the section were completely solid. This means that after solidification had begun at the centerline, feed metal had to travel only 1½ inches from the centerline to the solidification front instead of the full distance to the mold surface, as was the case in the sand casting.

At this point it is useful to define a term expressing quantitatively this difference in ease of feeding. It is desirable to have this term independent of absolute freezing time so that it can be used to compare different mold materials or different section sizes. In any event, the feeding of a casting becomes more difficult as soon as solid crystals are present at the centerline. The greater the percentage of the total solidification time during which these *centerline* crystals are present in a casting, the more difficult is the feeding.

We define [6] centerline feeding resistance as

$$\text{CFR} = \frac{\text{time during which crystals are forming at centerline}}{\text{total solidification time of casting}} \times 100. \tag{2-3}$$

This factor may be easily calculated from the start- and end-of-freeze curves:

$$\text{CFR} = \frac{\begin{array}{l}(\text{end of freeze time } at \text{ } centerline) \\ - (\text{start of freeze time } at \text{ } centerline)\end{array}}{\begin{array}{l}(\text{end of freeze time } at \text{ } centerline) \\ - (\text{start of freeze time } at \text{ } surface)\end{array}} \times 100. \tag{2-4}$$

For example, in the case under discussion, the steel poured in sand and in chilled molds would have the factors

$$\text{CFR}_{\text{sand}} = \frac{48 - 24}{48} = 50\% \text{ of the total solidification time,}$$

$$\text{CFR}_{\text{chill}} = \frac{10 - 8}{10} = 20\% \text{ of the total solidification time.}$$

From these data it may be concluded that chilled castings are inherently easier to feed than sand castings. This is found to be the case in practice, although for many years the reverse was thought to be true.

It may be noted that the centerline freezing ratio is really an index of the freezing bandwidth of the curves in Figs. 2–16 and 2–19. However, it is important to express bandwidth as a fraction or percentage, as is done in the CFR, and not as an absolute time value. In addition, since the most difficult region to feed is usually the centerline, the bandwidth percentage at this point is more important than at other regions. It should not be inferred that the presence of small amounts of solid at the centerline prohibits feeding; the start of solidification is merely a convenient reference point for calculation. It would be even better to use the time at which liquid cannot penetrate the crystal network, but this is more difficult to determine accurately.

TABLE 2–1

SUMMARY OF SOLIDIFICATION CHARACTERISTICS OF VARIOUS ALLOYS IN SAND AND CHILLED MOLDS

Material	Sand mold			Chill mold			Liquidus-solidus,		Heat of fusion, cal/cc	Thermal conductivity, cgs
	Center-line freeze time, min	Total freeze time, min	Centerline feeding resistance,* %	Center-line freeze time, min	Total freeze time, min	Centerline feeding resistance,* %	°F	°C		
0.6% carbon steel	26	48	54	1.7	9.0	19.0	120	67	525	0.076
18-8 steel (0.2% C max.)	15	43	35	0.9	9.6	9.4	30 (70)†	17 (39)	555	0.05
12% Cr steel (0.13% C)	16	42	38	0.9	9.7	9.3	40	22	545	0.06
Monel	40	63	64	0.8	9.6	25.0	95	53	600	0.06
Copper (99.8%)	0	32	<1	0.1	3.6	3.1	5	3	450	0.94
88-10-2 bronze	80	84	95	2.6	4.1	63.0	290	161	320	0.13
60-40 brass	10	38	26	0.1	4.0	2.5	20	12	300	0.28
Al, 8 Mg	63	69	91	3.4	5.2	65.0	115(260)†	64(140)	235	0.21
Al, 4.5 Cu	71	74	96	2.2	4.8	46.0	130(190)†	72(156)	260	0.33
Lead (99%)	5	30	17	0.8	4.8	17.0	15	8	68	0.083

* See Eqs. (2–3) and (2–4).

† In chilled castings, segregation produced the wider range shown in parentheses.

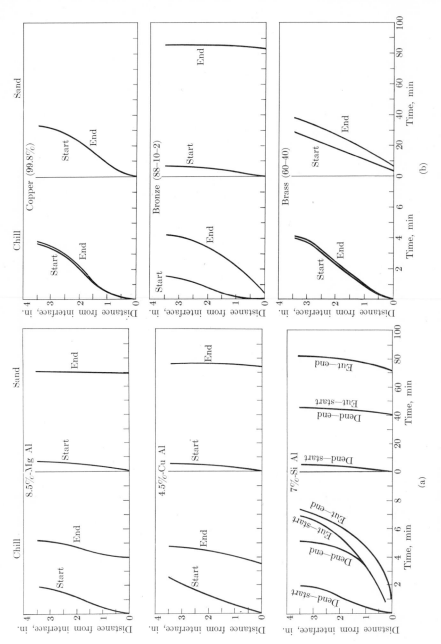

FIG. 2–20. Solidification of various alloys [12]. (a) Aluminum. (b) Copper. (*cont.*)

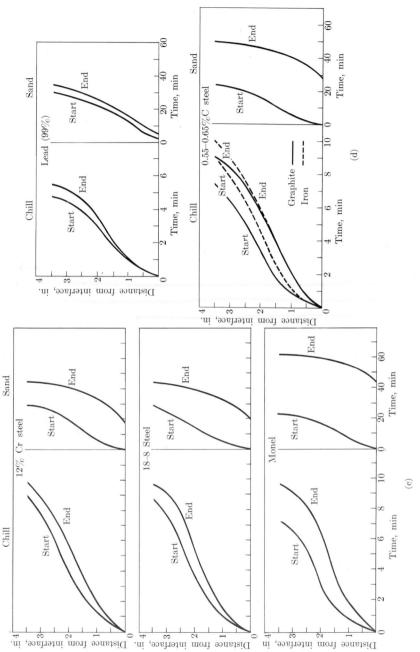

Fig. 2–20 (cont.). (c) Stainless steel. (d) Lead and steel [4].

(b) *Effect of alloy composition.* It is an experimental fact that some alloys are far easier to feed than others. In general, those with the smallest liquidus-to-solidus interval are the easiest to feed, while those with a long freezing range present difficulties. The freezing characteristics of a number of representative iron, copper, nickel, aluminum, and lead alloys are shown in Fig. 2–20 [4]. The centerline feeding resistance and liquidus and conductivity data are given in Table 2–1.

In general foundry experience, it has been found difficult to satisfactorily feed alloys which have a centerline feeding resistance of over 70%. The 88-10-2 bronze (CFR 95) behaves quite differently from the 60-40 brass (CFR 26), although both are copper-base materials.

The effect of chilling upon alloys is very helpful in reducing the CFR, as in the case of steel. The reductions are not directly proportional to the change obtained in the CFR for steel, however, because other factors, such as thermal conductivity, are important.

2–6 The rate of solidification. As mentioned earlier in this chapter, the proper design of a risering system depends upon the freezing pattern of the alloy (as evidenced by its centerline feeding resistance) and the freezing time of the riser relative to the casting. We now turn to the second of these considerations.

The object here is to develop a relationship, based solely upon the geometry of either a riser or a casting, that will permit determination of freezing times. It will then be possible to calculate, rather than having to determine experimentally, the minimum riser size which will provide a slower cooling rate than the casting and therefore serve as a liquid metal reservoir to supply feed metal during solidification of the casting (Chapter 3).

Since this calculation is rather complex, it is probably best to start by reviewing the elementary equation of heat flow for a steady-state condition. Figure 2–21 illustrates the heat flow through practically any furnace wall—cupola, open-hearth, or heat-treating furnace—after the temperatures of the inner and outer walls have been stabilized. Let us assume that the inside and outside walls have reached steady-state temperatures of 2400°F (1316°C) and 650°F (343°C), respectively.

The flux, V, of heat transmitted per unit time depends directly upon the temperature difference, the thermal conductivity, and the cross-sectional area of the wall, and inversely upon the distance between the hot and cold surfaces. If we assume that conductivity, K, is independent of temperature, then

$$J = K \frac{\Delta T}{\Delta x},$$ (2–5)

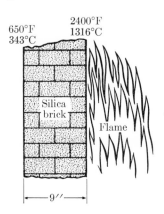

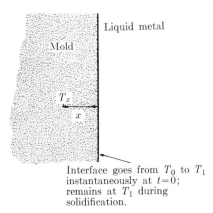

FIG. 2–21. Steady-state heat flow.   FIG. 2–22. Heat transfer in a sand mold.

where $J = $ Btu/hr·ft^2

$x = $ wall thickness in ft,

$T = $ temperature difference in °F,

$K = $ thermal conductivity in Btu·ft/°F·hr·ft^2, and

If K (for silica brick) is 1.06, then

$$J = 1.06 \frac{(2400 - 650)}{9/12} = 2480 \text{ Btu/hr·ft}^2$$

through a wall 9 inches thick. The quantity $\Delta T/\Delta x$ in this case is the *thermal gradient,* or the slope of a graph plotting temperature against distance in the brick. For constant conductivity, this graph is a straight line, and it is satisfactory to take the value (total temperature drop)/(total distance) as the thermal gradient.

Now let us review the differences between the steady-state condition just discussed and the case of pouring hot liquid metal into a cold mold. Here, we wish to find the heat flow through the mold, because it is obviously by this route that heat escapes, to permit solidification of the metal. If we can obtain a general expression for the temperature at any point x in the mold of Fig. 2–22, we shall be able to calculate the heat flow and thus the freezing time of the casting.

When a metal is poured into a mold, most of the heat is eventually absorbed by the mold itself, whereas in the steady-state case *all* of the heat from the inner surface is transmitted to the outer surface. For this reason, the specific heat of the mold material, C_{mold}, the density of the mold material, ρ_{mold}, and the thermal conductivity, K_{mold}, enter the heat-flow equation, and, for convenience, can be combined in one term:

$$\alpha_{\text{mold}} = \text{thermal diffusivity} = \frac{K_{\text{mold}}}{C_{\text{mold}}\rho_{\text{mold}}}.$$

Another important variable is the variation of the temperature T_x at any point in the mold, as a function of time. With these points in mind, let us review the general equation derived by Carslaw and Jaeger [3] and Ruddle [4].

Assume that the mold surface in Fig. 2–22 is raised instantaneously from an initial temperature T_0 to temperature T_1 at $t = 0$, and that it remains at T_1 during solidification. This situation is very closely approached in sand molds (Fig. 2–15). After a time t has elapsed, the temperature T at a distance x from the mold surface is given by [1]:

$$T_x = T_0 + (T_1 - T_0)\left(1 - \text{erf}\,\frac{x}{2\sqrt{\alpha t}}\right). \qquad (2\text{–}6)$$

The reader may not be familiar with the error function (erf), which may also be expressed as a convergent series:

$$\text{erf } u = \frac{2}{\sqrt{\pi}}\left(u - \frac{u^3}{3{\cdot}1!} + \frac{u^5}{5{\cdot}2!} - \frac{u^7}{7{\cdot}3!}\cdots\right). \qquad (2\text{–}7)$$

This complex expression is needed because of the changing thermal gradient in the sand as it heats. If we examine Eq. (2–6), we see that the temperature at point x at time t

(a) increases with increasing mold temperature T_0, temperature T_1 of the interface, and thermal diffusivity α, and

(b) decreases as the distance x from the interface increases.

If we are now to determine the rate of heat flow from the casting at the interface, we need to determine the thermal gradient, which may be calculated by differentiating Eq. (2–6):

$$\frac{dT}{dx} = \frac{d}{dx}\left[T_0 + (T_1 - T_0)\left(1 - \text{erf}\,\frac{x}{2\sqrt{\alpha t}}\right)\right]$$

$$= (T_1 - T_0)\frac{d}{dx}\left(\text{erf}\,\frac{x}{2\sqrt{\alpha t}}\right).$$

Substituting the series expansion for the error function and then differentiating, we obtain for the first two terms

$$\frac{d}{dx}\left(\text{erf}\,\frac{x}{2\sqrt{\alpha t}}\right) = \frac{2}{\sqrt{\pi}}\left(\frac{1}{2\sqrt{\alpha t}} - \frac{1}{2\sqrt{\alpha t}}\cdot\frac{3x^2}{3{\cdot}1!}\cdots\right).$$

Since we wish to obtain the gradient at the interface where $x = 0$, we write

$$\frac{d}{dx}\left(\text{erf}\,\frac{x}{2\sqrt{\alpha t}}\right) = \frac{1}{\sqrt{\pi \alpha t}},$$

and

$$\frac{dT}{dx} = -\,\frac{(T_1 - T_0)}{\sqrt{\pi \alpha t}} \quad \text{at}\quad x = 0. \tag{2–8}$$

Since the heat flow, J, per unit area per unit time at the interface is equal to the thermal conductivity times the gradient, as in Eq. (2–5), we have

$$J = \frac{K(T_1 - T_0)}{\sqrt{\pi \alpha t}}. \tag{2–9}$$

The minus sign for dT/dx in Eq. (2–8) is canceled in Eq. (2–9) due to the fact that the heat flow is down the thermal gradient.

Now the freezing time for a given casting or riser will simply depend upon removing a certain quantity of heat Q through a total area A at $x = 0$. Thus

$$Q = A \int_{t=0}^{t=t} \frac{dJ}{dt} = A\,\frac{2K(T_1 - T_0)\sqrt{t}}{\sqrt{\pi \alpha}}. \tag{2–10}$$

Let us determine the freezing time, t_s, of a large plate whose main faces have area A. The heat to be dissipated is the superheat above the freezing temperature and the latent heat of fusion. Let the pouring temperature be T_p and the solidification temperature be T_1 (the interface temperature); let C_{metal} be the specific heat of the liquid metal in Btu/°F·lb, L_{metal} the latent heat of fusion in Btu/lb, V the plate volume, and ρ_{metal} the density of the metal in lb/ft^3. Then the heat to be dissipated is

$$\rho_{\text{metal}} V \left[L_{\text{metal}} + C_{\text{metal}}(T_p - T_1)\right]. \tag{2–11}$$

Equating this result to Q above, we find for $t = t_s$:

$$\frac{V}{A} = \frac{2K_{\text{mold}}(T_1 - T_0)\sqrt{t_s}}{\rho_{\text{metal}}\sqrt{\pi \alpha_{\text{mold}}}[L_{\text{metal}} + C_{\text{metal}}(T_p - T_1)]}. \tag{2–12}$$

For a given combination of mold and metal conditions, we may solve for t_s and incorporate all constants into a single constant B, the *mold constant*. Then

$$t_s = B\left(\frac{V}{A}\right)^2, \tag{2–13}$$

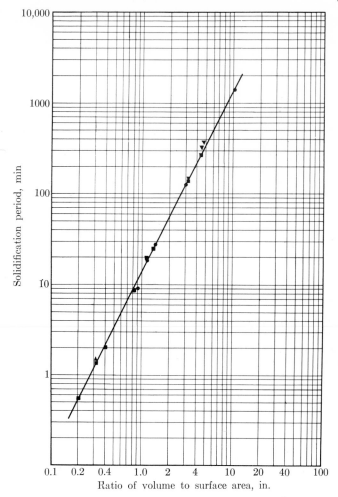

FIG. 2–23. Relation of freezing time to volume/area ratio [5].

where

$$B = \left\{ \frac{\rho_{\text{metal}} \sqrt{\pi \alpha_{\text{mold}}} [L_{\text{metal}} + C_{\text{metal}}(T_p - T_1)]}{2 K_{\text{mold}}(T_1 - T_0)} \right\}^2 .$$

Therefore, the freezing time is proportional to (volume/area)^2 of the casting or riser. Chvorinov [5] tested this relation for a wide variety of casting shapes and weights and found good agreement (Fig. 2–23). Correction factors which improve the agreement have been developed [4].

From these relations we may also determine t, the time required to freeze a given distance d, and the correlation of d with t. First, it should be noted that we can choose A as the area of one side if we take V as half the volume, and obtain the same result. Let us now assume a smooth freezing front for a plate; then, using the same concepts as in Eq. (2–11), we find that the heat to be removed in order to freeze a distance d is

$$(d)(A)(\rho_{\text{metal}})[L + C_{\text{metal}}(T_p - T_1)]. \tag{2–14}$$

Here A is the area of one side.

As before, the heat-transfer equation yielding Q', the new (lesser) amount of heat to be removed to freeze d, is

$$Q' = \frac{2AK}{\sqrt{\pi\alpha}}(T_1 - T_0)\sqrt{t}. \tag{2–15}$$

Equating and solving, we find

$$t = Bd^2, \tag{2–16}$$

where B is the *mold constant*. Therefore the thickness solidified varies with the square root of the elapsed time.

2–7 Correlation of analog computer data with Chvorinov formula and experimental results.

For a long period after the publication of the Chvorinov formula [5], there was considerable doubt as to its accuracy because of conflicting data from so-called "bleed-out" experiments. A number of investigators suggested that solidification time could be determined for a given mold geometry simply by pouring a series of castings of the given design and attempting to bleed or dump out the liquid metal after various time intervals. Data obtained in this manner yielded freezing times shorter than called for by the Chvorinov formula, and also gave the false impression that steel, for example, freezes with a single wave. No appreciable liquid + solid region was evident.

At about the same time, Paschkis [7] proposed that the electrical-analog method, which had been successful in other heat-transfer studies, be applied to the problem. In this method, the thermal circuit is represented by an analogous electrical circuit, since it has been found in many cases that electrical quantities are simpler to measure. The substitutions shown in Table 2–2 are made in the expressions for heat transfer. It is necessary, of course, to establish the proper relationships between thermal and electrical constants. This can be done by means of a few correlation experiments, where the solution to the thermal equation can be obtained from thermocouple measurements.

When the analog computer was used to develop freezing-time data, not only was good correlation with the Chvorinov formula obtained, but the

TABLE 2–2

Thermal system (casting or riser)	Electrical system (analog)
Thermal conductivity	Electrical conductivity
Thermal capacity	Electrical capacity
Temperature	Voltage
Heat flow	Current

freezing band, which did not emerge in the "bleed-out" experiments, was clearly evident. Thermal-analysis experiments [2] performed at about the same time also showed the freezing band, and hence indicated that the "bleed-out" experiments were in error. (The error was probably due to the difficulty of bleeding out the liquid from the dendritic network during the later stages of freezing; as a result, it falsely appeared that the metal was completely frozen at those stages.)

We have now satisfied the original objective of this chapter: to relate solidification phenomena to metal composition and mold geometry. We can now proceed to the practical application of what we have learned to the actual risering of a casting.

It should be remembered that in the theoretical calculations of freezing time, it was assumed that the casting solidifies with a smooth-freezing wavefront. The experimental data show that this is not true for most alloys; there are deviations which depend upon the composition of the metal. Because of this fact, it will not, in practice, be sufficient for the riser to freeze just a bit more slowly than the casting; instead an appreciable difference in freezing time, dependent upon the freezing pattern of the metal, may be required.

REFERENCES

1. H. F. Bishop, F. A. Brandt, and W. S. Pellini, "Solidification of Steel Against Sand and Chill Walls," Trans. A.F.S. 59, 435 (1951).
2. R. P. Dunphy and W. S. Pellini, "Solidification of Gray Iron in Sand Molds," Trans. A.F.S. 59, 427 (1951).
3. H. S. Carslaw and J. C. Jaeger, Conduction of Heat in Solids. London: University of London Press, 1947.
4. R. W. Ruddle, Solidification of Castings. The Institute of Metals (London), 1950.
5. N. Chvorinov, Proc. Inst. Brit. Found. 32, 229 (1938–9).
6. R. A. Flinn, "Quantitative Evaluation of the Susceptibility of Various Alloys to Shrinkage Defects," Trans. A.F.S. 64, 665 (1956).

7. V. Paschkis, "Studies in Solidification of Castings," *Trans. A.F.S.* **53**, 90–101 (1953).

8. A. H. Cottrell, *Theoretical Structural Metallurgy*, Chap. XIV. New York: St. Martin's Press, 1955.

9. R. Smoluchowski, ed., *Phase Transformations in Solids.* New York: Wiley, 1951.

10. J. Berry, V. Kondic, and G. Martin, "Solidification Times of Simple Shaped Castings in Sand Molds," *Trans. A.F.S.* **67**, 449–485 (1959).

11. W. A. Tiller, "Grain Size Control During Ingot Solidification," *J. Metals* Vol. 11, No. 8, 512 (1959).

12. J. L. Walker, *Liquid Metals and Solidification.* A.S.M., Cleveland, Ohio, 1958.

General References

Nucleation, solidification, growth:
Liquid Metals and Solidification. A.S.M., 1957
Transactions A.I.M.E., 1957. Contain many articles on the subject, notably those by B. Chambers, W. Tiller, J. W. Ruther, and J. L. Walker.
Phase diagrams and their interpretation: F. N. Rhines, *Phase Diagrams in Metallurgy.* New York: McGraw-Hill, 1956.

Problems

1. Explain why an ingot of deoxidized copper (Fig. 2–24) will exhibit a thin layer of randomly oriented crystals at the surface, followed by columnar crystals toward the center. Explain why an ingot of oxidized copper will show a region of randomly oriented crystals in the center. [*Hint:* Review the copper-oxygen phase diagram (*A.S.M. Metals Handbook*).]

Deoxidized copper   Oxidized copper

Figure 2–24

2. Explain the relative importance of thermal gradient and location of phase boundary lines for promoting constitutional supercooling.

3. How would the time required to freeze an ingot to a given distance d from the mold wall be affected by a change of the following variables: mold material; specific heat of the mold material; specific heat of the metal; latent heat of fusion of the metal; an allotropic transformation in the mold material during heating by the liquid metal.

Tell whether the effects will be linearly proportional, inversely proportional, etc.

4. Why does the theoretical assumption of a smooth-freezing wavefront not always hold in practice? How do you think deviations from the theoretical case will affect the riser size?

5. For the casting sketched in Fig. 2–25, show graphically how you would expect density, leak rate, and tensile strength to vary from cope to drag surface at locations A, B, C.

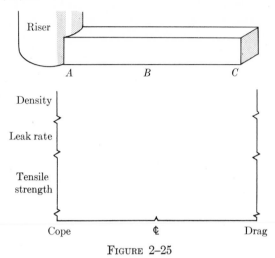

FIGURE 2–25

6. The heat Q flowing through area A in time t at the mold-metal interface is

$$Q = \frac{A \cdot 2K(T_1 - T_0)t}{\alpha},$$

where

$T_1 =$ interface temperature,

$T_0 =$ original mold temperature,

$\alpha =$ coefficient of thermal diffusivity of mold, and

$K =$ coefficient of thermal conductivity of mold.

Now if a mold is poured at temperature T_p, and if C_m is the specific heat of the metal and L is the latent heat of fusion per unit weight, determine the time required by the metal to solidify to a thickness d_t.

For a given set of mold and metal conditions, make an approximate plot of d as a function of t.

7. What would be the effect of changing the variables, mold material, pouring temperature, specific heat of the mold material, latent heat of fusion of the metal, upon the time required by an ingot (an ideal pure metal) to solidify to a given distance d from the mold wall. Give a quantitative answer, that is, say whether the time would increase linearly, inversely, proportional to the square of a given constant, etc. Justify your answer in each case.

CHAPTER 3

RISER DESIGN AND PLACEMENT

3–1 General remarks. We will now show how the theory of Chapter 2 and supplementary material can be applied to the actual quantitative design of risers for complex castings.

Risering is, of course, a process designed to prevent the formation of shrinkage voids in the casting upon solidification. To indicate the proportion of solidification shrinkage in the total contraction of a casting, let us review the relevant figures (in percent) for plain carbon steel during the three stages of contraction illustrated in Fig. 3–1:

(a) During the liquid state (contraction in the liquid metal), approximately 1.6% by volume per 100°F (56°C);

(b) during the transformation from the liquid to the solid state, about 3.0% by volume (values for other alloys are given in Table 3–1);

(c) during the solid state, about 7.2% by volume, from the solidification range to 70°F (39°C).

The point to recognize is that contraction (b) can take place at constant temperature or over a narrow range, and is the result of a density change accompanying the transformation from liquid to solid. Visualize a one-inch cube of liquid steel at the solidification temperature, as shown

TABLE 3–1

SOLIDIFICATION CONTRACTION FOR VARIOUS
CAST METALS

Metal	Percent volumetric solidification contraction
Carbon steel	2.5 to 3%
1% Carbon steel	4
White iron	4 to 5.5
Gray iron	Expansion to 2.5
Copper	4.9
70% Cu-30% Zn	4.5
90% Cu-10% Al	4
Aluminum	6.6
Al-4.5% Cu	6.3
Al-12% Si	3.8
Magnesium	4.2
Zinc	6.5

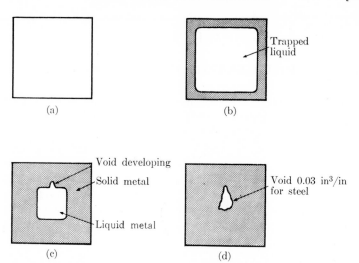

FIG. 3–1. Development of shrinkage void in a cubic casting. (a) Liquid.
(b) Liquid + solid. (c) Liquid + solid. (d) Solid.

in Fig. 3–1. A thin outer shell will freeze at constant temperature, and there will then be entrapped within the fixed outer dimensions a volume of liquid whose subsequent transformation to the denser solid will produce shrinkage of type (b). In castings of irregular sections it is also possible for the shrinkage of stage (a) to add to the voids if the gates are thin and freeze off before the casting reaches the freezing range. Note that shrinkage voids are not produced by contraction of type (c) which occurs in the solidified metal. (This type can, however, result in serious stresses, tears, and cracks, as discussed in Chapter 6.)

The problem at hand, shrinkage porosity, can be solved by controlling the solidification pattern and thermal gradients, so that the voids are produced outside the body of the casting proper. It is evident that to develop the best engineering properties in a given casting, the shrinkage voids must either be eliminated or isolated at a location of low stress.

We shall see that to produce a sound casting, the riser or reservoir of liquid metal which is to compensate for the shrinkage must satisfy two independent requirements.

(a) *Riser size.* If the riser is to supply liquid metal to feed the casting shrinkage, it must freeze after the casting (except in certain gray irons discussed at the end of this chapter). When the mold material surrounding the riser and the casting is the same, then the ratio (volume/area)^2 of the riser must exceed that of the casting.*

---

* Unless an external source of heat such as an exothermic compound or electric arc is placed over the riser, or the casting is chilled.

(b) *Riser placement.* For an alloy with high centerline feeding resistance, a casting will require a closer spacing of risers than for other alloys. In other words, the *effective feeding distance* of a riser in a "wide freezing-band alloy" is smaller than in a "narrow-band alloy."

It is evident that because of the differences among alloys in solidification patterns, the excess of riser-freezing time over casting-freezing time and the variation in riser placement may vary widely. However, the fundamental ideas which have been evolved in Chapter 2 provide a good background for developing and applying the experimental data.

We shall proceed now to discuss the construction of several graphs used to calculate the riser size and then take up the calculation of riser placement. Since most of the published work to date has dealt with steel and ductile iron, the discussions will necessarily be limited to these materials. Special attention to gray iron, brass, bronze, and aluminum is given, however, at the end of the chapter.

3–2 Risering curves for steel. Riser size is determined by two factors: first, the freezing time of the riser must exceed that of the casting, at least to some extent, and second, the riser must supply sufficient feed metal to compensate for the liquid-to-solid shrinkage. Caine [1] has evaluated these requirements as shown by the risering curve of Fig. 3–2(a).* In the figure, the x-axis represents the ratio of the freezing time of the riser to that of the casting, which, as mentioned above, must always be greater than 1 : 1 to obtain a sound casting. As an index of freezing time, Caine uses the ratio (surface area/volume), which is inversely related to the V^2/A^2 ratio of Chvorinov, developed in Chapter 2. Caine's reasoning is that heat dissipation is a function of the surface area of the casting, while the heat content is a function of volume. It is assumed intuitively that the linear relation A/V determines cooling *rate* and hence is inversely related to freezing time. While the Chvorinov relationship has a firmer theoretical basis, the simpler Caine ratio apparently falls within the limit of error, considering the many other factors involved.

The y-axis represents the volume ratio of riser to casting. Here it is postulated that the riser will be required to furnish feed metal to compensate at least for the liquid-to-solid shrinkage, which is approximately 3% by volume for steel. The riser size, therefore, must be greater than 3% of the casting volume. This limiting case of high yield, i.e., large casting/riser ratio, is reached only for extremely thin plates of very large surface/volume ratio. For example, a plate of dimensions $10 \times 10 \times 0.1$

---

* The curve in Fig. 3–2 (a) does not include any safety factor, which can be allowed for by increasing the constant c in the equation by whatever amount is deemed advisable. For example, a factor of 5% should be enough for steel.

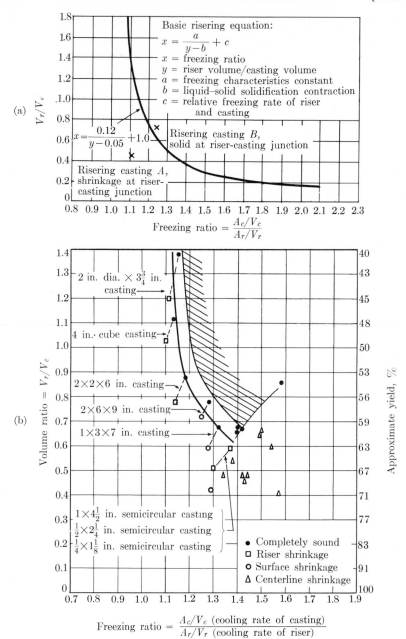

FIG. 3-2. (a) Risering curve for 0.3% C steel castings and the basic risering equation used to calculate riser dimensions [1]. (b) Risering curve for ductile iron (total carbon 3.6%, milicon 2.5%) [15]. Cross latched region is the most efficient area for risering to complete soundness; side blind risers, clamped green-sand molds gated into risers.

in. has an A/V-ratio of

$$\frac{100 + 100 + 4(10 \times 0.1)}{10 \times 10 \times 0.1} = 20.4.$$

This volume would require 0.3 in^3 of feed metal. A cube riser of 0.5 in. would have the ratio

$$\frac{A}{V} = \frac{0.5 \times 0.5 \times 6}{0.5 \times 0.5 \times 0.5} = \frac{1.5}{0.125} = 12,$$

or a much slower cooling rate than the plate. However, the amount of feed metal, 0.125 in^3, would be insufficient even if it were available. For this reason, castings with high surface area/volume ratios require larger risers than called for by cooling-rate considerations alone, as shown by the right-hand portion of the graph in Fig. 3–2.

On the other hand, as the casting becomes compact and its cooling rate decreases, there is always adequate feed metal in the riser, since, to cool at a sufficiently slow rate, the riser must be both compact and larger than the casting in all dimensions. Consider, for example, the feeding of a 4-in. cube:

$$\frac{A}{V} = \frac{4 \times 4 \times 6}{4 \times 4 \times 4} = \frac{96}{64} = 1.5.$$

A cylindrical riser 4.5 in. in diameter and 4.5 in. high will be needed to produce a sound casting, as shown by the following calculation. Using the most compact cylinder (diameter = height), so that $A/V = 6/d = 1.33$, we find that its A/V-ratio is

$$\frac{A/V_{\text{casting}}}{A/V_{\text{riser}}} = \frac{1.5}{1.33} = 1.13,$$

and its volume ratio is

$$\frac{V_{\text{riser}}}{V_{\text{casting}}} = \frac{71.5}{64.0} = 1.11.$$

These coordinates satisfy the graph of Fig. 3–2. Improvement in yield can be obtained by the use of chills, which can increase the effective surface area of cooling by a factor of five. Additional risering-curve data (for ductile cast iron) are given in Fig. 3–2(b).

3–3 **Naval Research Laboratory method of riser calculation.** After several years of successful use of the Caine risering calculation, the NRL group [2] worked out a new and simplified procedure (illustrated in Figs. 3–3, 3–4, and 3–5) which has the advantage of eliminating trial-and-error calculations. The method is based on the observation that the ratio

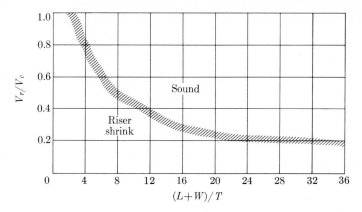

FIG. 3–3. Risering curve for steel (approximately 0.2% C to 0.5% C) [2].

(riser volume/casting volume) must be greater for chunky castings because of their relatively low surface/volume ratios. (This requirement is also a characteristic of the Caine relationship.) A major simplification is obtained, however, by the device of introducing the shape factor $(L + W)/T$ to replace the tedious calculation of surface/volume ratios for complex castings. The length L, width W, and thickness T are computed by using the maximum dimensions of the *parent* section of the casting. For example, if a casting has a number of small appendages, then these are not used in the calculation of the shape factor; however, they do enter into the calculation of the casting volume in a special way, as we shall discuss in Section 3–7.

The NRL method may be simply illustrated by the following practical example. The most economical riser size for a steel casting of dimensions $5 \times 10 \times 2$ in. is calculated by first computing the shape factor $(L + W)/T$, which for this casting is $(10 + 5)/2 = 7.5$. Then, from Fig. 3–3, the ratio (riser volume/casting volume), or R_V/C_V, is found to be 0.55. Since the volume of this casting is 100 in³, the riser volume required is 0.55×100 in³ $= 55$ in³. From Fig. 3–4 it is seen that this volume is provided by a riser 4.5 in. in diameter and 3.5 in. high.

However, it can be shown empirically that the most economical h/d ratio for risers attached to the side of the casting is unity. On the other hand, the most economical h/d ratio for top risers is one-half, the lower curve on each of the charts. Therefore, if the casting under consideration is to be fed by a side riser, the best riser dimensions are: diameter 4.25 in. and height 4.25 in. If, however, a top riser is employed, it should be 5.25 in. in diameter and 2.63 in. high or, more practically, 5.0 in. in diameter and 2.75 in. high.

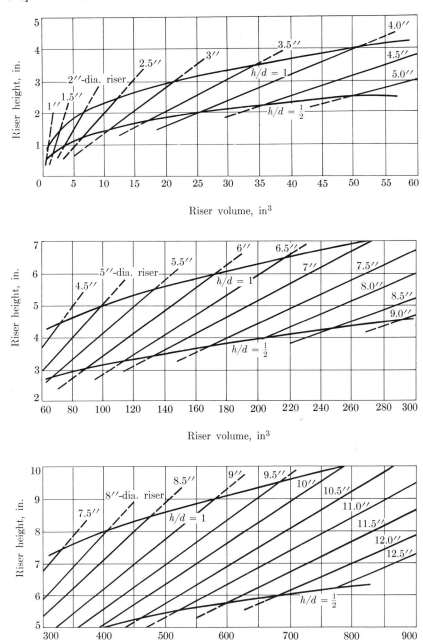

Fig. 3-4. Chart for conversion of required riser volume to riser dimensions.

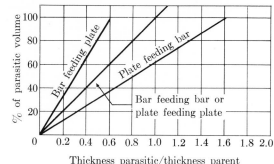

FIG. 3–5. Naval Research Laboratory risering chart for irregular sections. (a) Percentages of parasitic volume to be added to parent volume to determine riser volume.

For a circular plate 10 in. in diameter and 2 in. thick, the $(L + W)/T$ shape factor is $(10 + 10)/2 = 10$, but the actual volume of 157 in^3 is used in the volume calculation for the proper amount of feed metal. For a cylinder 4 in. in diameter and 10 in. long, $(L + W)/T = (10 + 4)/4 = 3.5$, and again the actual volume is employed.

It has been demonstrated in practice that both the NRL and the Caine curves give approximately the same riser dimensions for simple shapes. The NRL method, however, has been further extended to include calculations for complex shapes such as those shown in Fig. 3–6. When ribs or other appendages are thin, they do not appreciably increase the freezing time of the main portion of the casting, and therefore only a small increase of liquid metal is needed in the riser to allow for the appendages. As the appendage becomes heavier, the riser must be increased considerably to ensure that the freezing time of the riser is sufficiently longer than that of the casting. Very thin fins can be used to reduce the cooling time; however, the effect of such an arrangement is difficult to calculate and is not taken into account in Fig. 3–5.

The calculation of additional riser volume in terms of the percentage of liquid metal to be added to the riser volume is indicated in Fig. 3–5. All appendages can be considered as approximations of bars or plates.* Because of the greater surface area, a bar will have a greater cooling rate than a plate of the same thickness. This is evident in the case of a plate feeding a bar: a plate 1 in. thick is equivalent to a bar 1.6 in. thick. At this size both have the same cooling rate, and the bar must be treated as part of the parent casting. However, when a plate fed by a bar

---

* A casting is considered to be a plate when the width of its cross section is greater than three times the height. When the width is less than three times the height, the casting is considered to be a bar.

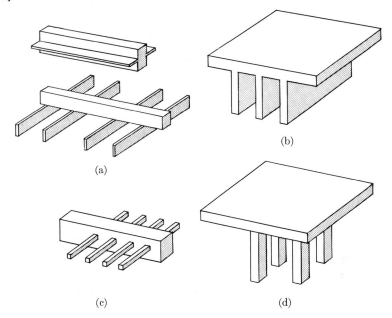

FIG. 3–6. Examples of casting conditions of different shapes and thicknesses [8]. (a) Bar–plate. (b) Plate–plate. (c) Bar–bar. (d) Plate–bar.

reaches 60% of the bar thickness, then it must no longer be treated as parasitic.

To illustrate the calculation, let us assume that a projection of cross section 1 in² and 4 in. long is added to the plate of dimensions $5 \times 10 \times 2$ in. discussed earlier. This addition does not enter into the calculation of the $(L + W)/T$ factor, which remains at 7.5. The volume of the parasitic bar is 4 in³, and from Fig. 3–5 it is evident that 30% of the parasite volume, or 1.2 in³, should be added to the parent volume. A riser volume of 55.6 in³ is now required.

Hollow cylindrical shapes such as bushings also present a special case. The heat flow from the center core is restricted, and the casting as a whole has a lower cooling rate than a plate of the same cross section. The simplest approach is to consider the shape as a plate, but to correct for an "effective plate thickness" (as indicated by experimental observations) as follows.

Let T be the true wall thickness and T_e the effective plate thickness. Then the following approximations may be used:

Core diameter	$\frac{1}{2}T$	$1T$	$2T$	$4T$	(flat plate)
Correction factor (k)	1.17	1.14	1.02	1.0	(1.0)

$$T_e = kT$$

As an illustration, let us consider the computation for a bushing whose dimensions are: outside diameter 12 in., inside diameter 4 in., and height 12 in. From the above table, the thickness of an equivalent plate is 4.56 in. The shape factor is

$$\frac{8\pi \text{ (circumference)} + 12}{4.56} = 8.1,$$

$$\frac{R_V}{C_V} = 0.47,$$

and the actual casting and riser volumes are

$$C_V = (6^2 - 2^2)12\pi = 1205 \text{ in}^3,$$

$$R_V = 567 \text{ in}^3.$$

Hence, according to the chart (Fig. 3–5), the riser should be $11\frac{1}{2}$ in. in diameter and $5\frac{1}{2}$ in. high.

3–4 Feeding distance. All of the preceding calculations are based upon the assumption that the riser or risers are placed within the effective feeding distances of the casting sections requiring liquid metal. For example, from the curve it is possible to calculate the dimensions of a riser designed to feed a bar 1 in. in diameter and 21 in. long, but obviously, because of the long feeding path, the resulting section, having only a single riser at one end, would not be sound. Let us now proceed to calculate effective feeding distances from the available experimental data. It will be recalled from the discussions of Chapter 2 that each combination of alloy and mold material exhibits its own characteristic centerline feeding resistance. It is necessary to determine experimentally the effective feeding distance for each combination until sufficient data are compiled to permit the derivation of a general relationship. The most detailed experimental results for cast steel have been published by the Naval Research Laboratory [3, 4].

In the determination of feeding distance, it is assumed, for simplicity, that any casting can be divided into plate, bar, and cubical (or spherical) sections. The cubical or spherical sections offer no problem of feeding distance because the riser can be placed near the location to be fed. Hence, if we can develop feeding distance data for bar and plate sections, we shall be able to position risers properly for any casting.

The plate and bar sections require individual attention because in the plate we encounter dendritic growth proceeding from two principal walls, while in the bar four walls are involved. It is necessary, then, to pour liquid metal and observe the feeding distances from risers in bars and in plates of different thicknesses.

(a) *Feeding distance in bar sections.* The NRL data for 4×4 in. bars of different lengths are presented in Fig. 3–7. Two types of observations were made: first, temperature measurements were performed at various centerline locations in the bars during cooling, and then the bars were inspected radiographically. In all cases the radiographic inspection disclosed a sound section near the riser and another at the far end of the bar. The 12-in. bar was completely sound, but both the 16-in. and the 24-in. bars exhibited centerline shrinkage. There was some scatter in the results, as shown by the overlapping of the crosshatched areas in the figure. It is interesting to note that the shrinkage zone approached the riser more closely than the far end of the bar. The thermal data explain this observation quite convincingly.

The thermocouple readings are then used to develop a plot of temperature distribution at selected time intervals. This method is similar to that employed in graphs of solidification (Chapter 2). For clarity, a few selected lines from Fig. 3–7(b) are reproduced in Fig. 3–8. In the feeding of a given centerline location, the thermal gradient is most important for the period of time during which the last stage of solidification (solidus) occurs. After this time interval has passed, nothing can be done to fill in shrinkage voids which may have resulted during solidification. During the last stage of solidification, the dendritic growth in the area, and hence the centerline feeding resistance, will be at a maximum.

To determine the thermal gradient at a given station during the last stage, a tangent is drawn to the particular curve of distance versus temperature which passes through the solidus at that station. Note that the gradient for each station is determined at a different elapsed time but at the same temperature (the solidus). By comparing the thermal-gradient data derived in this way with the soundness data, one can see that in all cases a thermal gradient of greater than $6°F/in.$ is required to allow proper feeding. The 16- and 24-in. bars exhibit gradients of 0 to $2°F/in.$ at the center regions, indicating that these sections complete their freezing, and so block off feed metal at about the same time.

The gradient data also explain why the sound region encountered at the end of the bar away from the riser is longer than that at the riser. The faster cooling rate provided by the end face causes this section to freeze well in advance of the center portion, thus drawing feed metal from the center. This metal, of course, is replaced by riser metal. The cooling effect of the end face provides a greater thermal gradient over a greater distance than does the riser.

After experimenting with bars from 2 in. to 8 in. in cross section, the Naval Research Laboratory derived the following general relationship *for bars in this range:* effective feeding distance $D_{max} = 6\sqrt{T}$, where T is the bar thickness and D is measured from the *edge* of the riser.

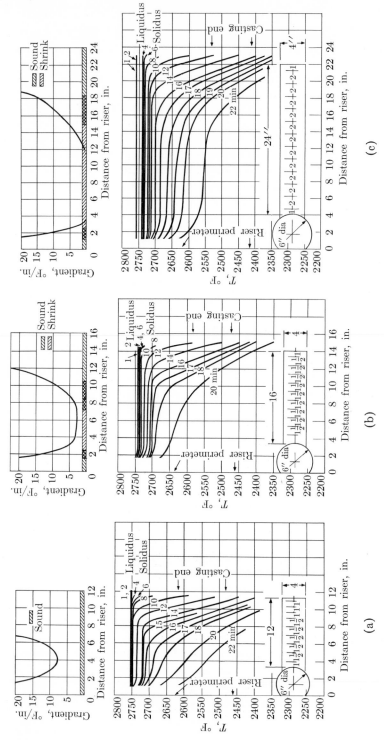

Fig. 3–7. Solidification of steel bars of cross section 4 in. [3].

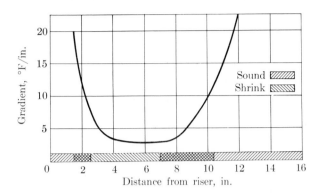

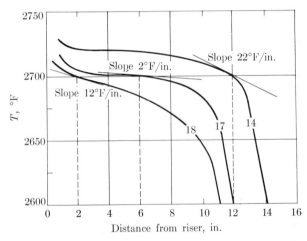

Fig. 3–8. Calculation of thermal gradient from Fig. 3–7(b).

When the bar is of greater length than the feeding distance of a riser, it is necessary to use additional risers. It should be emphasized that the feeding distance of a riser can be applied in any direction. For example, in the 4×4 in. bar already discussed, if a riser 8 in. in diameter were centrally located, a sound 32-in. bar could be produced (12-in. feeding distance in each direction from the riser *edge*). If a still greater length is required, it is necessary to place two risers on the bar. In this case, the feeding distance for metal *between* the risers is not $6\sqrt{T}$ from each riser edge, but is $1.2T$ because there is no cold end face *between* the risers, Fig. 3–9.

The individual contributions of the riser and of the cold face at the end of the bar to the production of sound regions are shown in Fig. 3–9. These contributions add up to the maximum length of sound bar that can

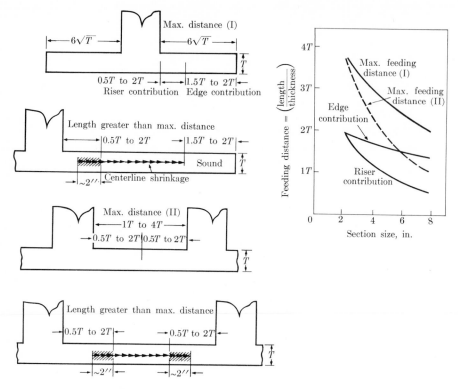

Fig. 3–9. Feeding relationships in bars [8].

be obtained; for example, in a 4-in. bar, the riser contribution is $1.2T$, that of the edge is $1.8T$, and hence the total contribution is $3T$ or 12 in.

(b) *Feeding distance in plate sections.* The feeding distances in plates have been established by a series of experiments [4] similar to those for bars, using plates of thickness $\frac{1}{2}$ in. to 4 in. As predicted, a given riser has a greater feeding distance in a plate section because dendritic growth occurs only from two principal walls, and therefore less resistance is offered to the transport of feed metal. A thermal gradient as low as $1°F/in$. can be tolerated for a horizontal distance equal to the thickness of the plate, in comparison to the requirement of $6°F/in$. minimum for bars. The feeding distance for plates 1 to 4 in. in thickness is $4.5T$ when both riser and cold end-wall gradients are taken into account. The riser gradient prevails for a shorter portion of the total distance ($2T$) than the end-wall gradient ($2.5T$), just as for bars. Therefore, when the feeding distance between risers is calculated, each riser feeds a distance equal to $2T$, not $4.5T$. These relationships are summarized in Fig. 3–10.

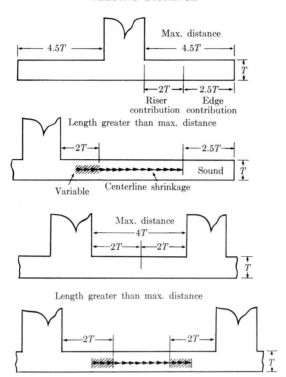

FIG. 3–10. Summary of end wall and riser gradients in plates [8].

(c) *Feeding distance in complex sections.* In calculating riser size it was necessary to develop relationships among combinations of bars and plates (parent versus parasite), and the feeding distance problem requires similar attention. When a heavy plate joins a light plate for its full width (Fig. 3–11), the heavy plate is more effective than a riser. This follows from the fact that the riser has only a limited attachment, at its periphery, to the plate, whereas the heavy plate feeds throughout its length. By a series of systematic experiments it was established [8] that the feeding distance in the thin plate is $3.5T_H$, where T_H is the thickness of the *heavy* section. Therefore a 1 in. plate can be fed by an adjoining 4-in. plate section for 14 in. instead of for only 4.5 in.

The feeding distance of a riser over the heavy section, however, is reduced to $D_H = (T_H - T_L) + 4.5$, where D_H is the feeding distance of the riser in the heavy section and T_H and T_L are the thicknesses of the heavy and light sections, respectively. More elaborate modifications have been developed for intermediate sections, for which we find the relation $D_M = 3.5T_H - T_L$, where D_M is the feeding distance in the middle section. When a very light section, e.g., a $\frac{1}{8}$-in. plate, is attached to a heavy

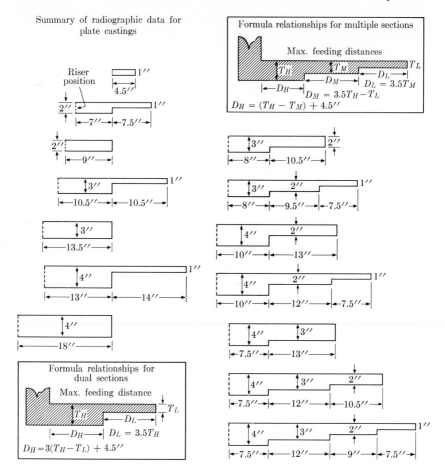

FIG. 3–11. Feeding distances in varying plate sections [8].

(3-in.) section, the feeding distance in the heavy section is not reduced, since the light section accelerates rather than slows down the end-plate cooling.

The case of a light section bar attached perpendicularly to a heavy plate has not been investigated, but it may be expected that the feeding distance in the bar will be the same as for a well-risered bar.

3–5 Other effects of complex sections and designs. From the many examples already discussed, it should be apparent that any alteration in a section will lead to a change in cooling rate. Such alterations are of two general types: (a) changes in direction (L-shaped versus straight bar) and (b) presence of intersections.

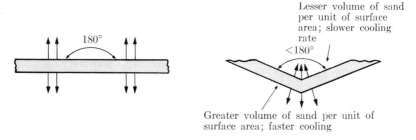

FIG. 3–12. Effect of change in direction, uniform section [5].

FIG. 3–13. T, X, V, and other intersections [5]. (Regions marked by crosses are "hot spots.")

(a) *Effects of changes in direction.* One effect of a change in direction of the uniform section illustrated in Fig. 3–12 is the decrease in the volume of sand per unit of surface area which occurs on the side where the included angle is less than 180°. When a condition of this type is encountered, the casting may develop shrinkage because of the retarded cooling, i.e., the metal on both sides of the hot spot tends to freeze and wall off the zone from feed metal.

Other difficulties, such as hot tearing and development of residual stresses, are also introduced. These will be discussed in Chapter 6.

(b) *Effects of intersections.* Intersections provide a striking case of retarded cooling rate, as illustrated in Fig. 3–13. There are three ways of coping with this problem.

(i) It is often effective to chill the side of the section where the re-entrant angle is located, although this technique increases the molding cost. Also, the chilling may be overdone, and there is danger of choking off the feed metal so that hot tearing may result at the edge of the chill.

(ii) Redesigning to lighten the metal section at the point of intersection is often more desirable than method (i). (Full details are given in reference 5.)

(iii) Another possibility is to mold a thin fin across the valley of the intersection. This increases the surface area of the metal and therefore the cooling rate.

3–6 Effect of chills. It was pointed out in Chapter 2 that the substitution of metal chills for all or a portion of the sand-mold surface has two pronounced effects. Because the cooling rate of the casting is increased, the start and end portions of the freezing waves are closer together in moving from the surface to the center of the casting. This means that the mushy-liquid + dendritic-solid region is narrowed and the centerline feeding resistance is reduced.

Chills, therefore, have two distinct functions in producing sound castings. First, the feeding distance in a given riser-casting combination is increased because the centerline feeding resistance is lowered. Unfortunately, no data are yet available to illustrate this effect quantitatively. One outstanding example, however, may be cited. The Griffin Wheel Laboratories have developed a process for making a railroad car wheel (33 in. in diameter) of 0.7% carbon steel, in a graphite mold. The entire rim section (100 in. in circumference), as well as most of the plate or web section, is fed from four risers, so that a feeding distance of over 12 in. per riser is indicated. Since the rim can be approximated by a 2-in. bar, we have here a feeding distance of $6T$ as compared with the distance $2T$ expected in a sand mold. This difference arises from the decrease in centerline feeding resistance for 0.6% carbon steel which drops from 54% to 19% (sand versus chilled) as calculated in Chapter 2.

Before proceeding further we must emphasize that the effects of chills just discussed were for *uniformly chilled castings*. It is found that even with uniform chilling, which causes more rapid freezing than sand does, the freezing distance is improved in almost all cases. Only when the thermal conductivity of the metal is much greater than that of the chill, e.g., for copper, is this method of chilling less effective.

Chills are far more commonly used to obtain a steeper thermal gradient in the longitudinal direction. In a plate or bar casting where it is necessary to place several risers on top of the section (Fig. 3–14), the feeding distance of the risers may be markedly increased and their number reduced by chilling. Consider first a chill placed at the end of a plate or bar section. Its effect on feeding distance is relatively minor,

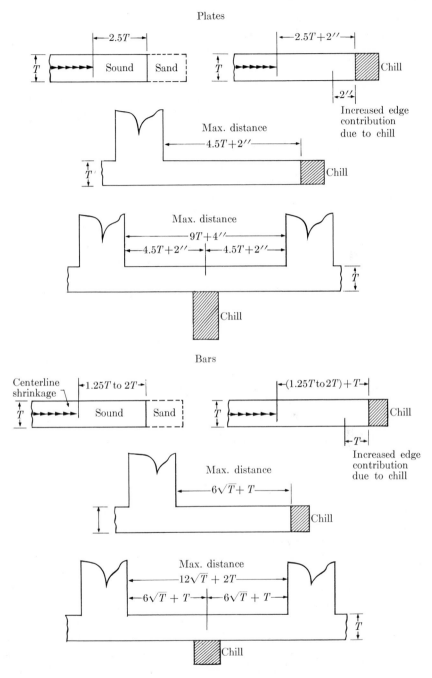

FIG. 3–14. Combination of chills and risers for maximum feeding distance [8].

since a strong gradient is already present, caused by the sand-metal interface. Accordingly, the feeding distance is increased only by a factor of T (the section thickness) for a bar, and by a factor of $2T$ for a plate, (Fig. 3–14).

By contrast, let us consider the effect of a chill placed between risers. [With merely a chill in the drag (bottom) surface, the effect upon the thermal gradient is the same as if an end chill were placed in the mold (Fig. 3–14).] The feeding distance of each riser is now $4.5\,T + 2$ in., and for a bar, it is $6\sqrt{T} + T$. Thus, for example, in a 1-in. plate, the distance between risers can now be 13 in. instead of 4 in. This effect is of obvious importance in all alloys when large plate sections, such as ball-mill liners, are to be fed by a number of risers. In the selection of chills for this purpose, we are guided by the results of thermal studies showing that chills of thickness and width equal to the plate are adequate. No increase in cooling rate during the critical solidification period is obtained with water-cooled copper chills or heavier metal chills.

3–7 Application of risering principles to complex castings. To illustrate the calculation of riser size and placement let us consider the casting of a bearing housing shown in Fig. 3–15(a). We should recognize first of all that this casting is made up essentially of two plates, the base and the cylindrical section (unrolled). Secondly, it will not be possible to feed this casting with one riser because of the feeding distance involved. Hence we decide to riser the casting from two locations as shown in Fig. 3–15(b).

The base and half the cylinder may be considered as a plate $11 \times 5.5 \times 1$ in. with an appendage $8 \times 5 \times 1$ in. The shape factor is

$$\frac{11 + 5.5}{1} = 16.5,$$

whence we obtain a risering volume factor of 0.25. The volume of this shape is 105 in³, and therefore a riser 3.5 in. in diameter and 2.7 in. high is required. Note that the volume of the base is taken at full value because it is as thick as the cylinder. The half cylinder and boss away from the base have the same shape factor (16.5). The volume of the boss is taken at full value which, combined with the half cylinder, equals 72 in³. Thus a riser 3 in. in diameter and 2.6 in. high is needed.

Since the feeding distance of the risers is not sufficient, chills are employed as shown in the sketch. The feeding distances from the riser to the end of the base and to the bottom of the cylindrical section are close to the maximum, and a pilot casting should be made to determine whether additional chills are needed.

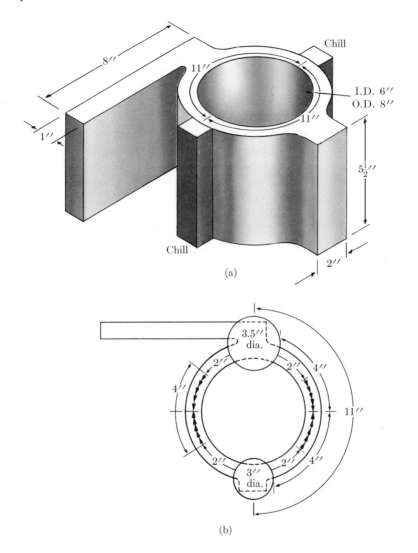

(a)

(b)

FIGURE 3–15

3–8 Risering of gray iron. The risering of gray iron has been reserved for special attention, since it is a complex problem having some features in common with steel, whereas others are completely different. According to the thermal-analysis data given in Chapter 2, solidification of a 3% carbon gray iron takes place by two distinct mechanisms:

(1) Liquidus to eutectic; separation of austenitic dendrites.
(2) Eutectic; concurrent precipitation of graphite and austenite.

Reaction (1) is similar to that for steel, i.e., the precipitation of solid austenite involves the formation of a denser phase than that of the liquid, and therefore feed metal is required from a riser at this time. Also, as shown by the thermal analysis data of Chapter 2, reaction (1) practically reaches completion before the inception of reaction (2). During reaction (1), then, liquid is transported from the riser to the casting.

On the other hand, reaction (2) involves an expansion, since the density of the solid eutectic mixture is less than that of the parent liquid. During this reaction, metal flows back into the riser, a process known as "purging" [6]. If the riser is small and has frozen over at the top, great pressure will be generated within the casting and the walls will actually be bowed outward. However, if the riser is open to the atmosphere, the pressure is relieved by metal flow. This can be of major importance in determining dimensional accuracy.

The relative shrinkage due to reaction (1) and the expansion of reaction (2) cancel each other [7] when about 1.5% graphitic carbon is precipitated in the eutectic. This carbon value is related to the solubility of carbon in the austenite, which in turn is affected by silicon. The percentage of eutectic graphite can be expressed as

$$\text{Eutectic graphite} = \text{total carbon} - 2 + 0.1\,(\%\text{Si}).$$

When this value is equal to 1.5% carbon or more, no risering is required. Below this value risers are needed in proportion to the decrease in the percentage of eutectic graphite. Gray iron is, therefore, the only common exception to the rule that the riser must freeze after the casting. The riser need stay liquid only long enough to meet part of the early casting demand for metal to compensate for the liquid $\rightarrow$ austenite reaction. At the end of this reaction, the eutectic reaction will produce an expansion and will "purge" part of this riser metal back to the risers.

3–9 Risering of brass, aluminum, and magnesium. A search of the cast-metals literature will disclose fully as many articles dealing with these nonferrous materials as with steel. Why then has so much space been given here to steel? The key lies in the fact that the data for steel (and also for ductile iron and gray iron) are quantitative, whereas the other data are largely qualitative and descriptive.

We may, however, predict something about the feeding behavior of a number of other alloys from the centerline feeding-resistance data of Chapter 2. In sand molds we expect that alloys with a centerline feeding resistance similar to that of steel will have similar risering characteristics, whereas those with high resistance will be more difficult to feed. We can predict relationships like those shown in Table 3–2.

TABLE 3-2

CENTERLINE FEEDING RESISTANCE OF VARIOUS ALLOYS

Alloy	Centerline feeding resistance	
18-8 steel (0.2%C)	35	
12% chromium steel	38	less than that
Copper	1	of 0.6%C steel
60-40 brass	26	
Lead	17	
0.6%C steel	54	similar to
Monel	64	that of steel
88-10-2 bronze	95	
Aluminum, 8% magnesium alloy	91	greater than
Aluminum, 4.5% copper	96	that of steel

Recent work at the Naval Research Laboratory indicates that these predictions are correct. The feeding distance for manganese bronze (similar to 60-40 brass) in plates is 5.5 T versus 4.5 T for steel, where T is the thickness of the plate. Furthermore, in 88-10-2 bronze, shrinkage is encountered along the full length of the bar.

In general, then, it appears that the risering data for steel may be applied to all alloys whose centerline feeding resistance is lower than that of steel. For alloys with great resistance, local chilling is advisable at those portions of the casting for which maximum strength is indicated. For example, it has been shown for 85-5-5-5 bronze that for 2×2 in. bars and plates 1 in. thick, chilling is required to produce a gradient of over 60°F/in., which is the minimum gradient that will ensure soundness in this material [10].

REFERENCES

1. J. B. CAINE, "Risering Castings," *Trans. A.F.S.* **57**, 66–76 (1949).

2. H. F. BISHOP, E. T. MYSKOWSKI, and W. S. PELLINI, "A Simplified Method for Determining Riser Dimensions," *Trans. A.F.S.* **63**, 271–281 (1955).

3. H. F. BISHOP, E. T. MYSKOWSKI, and W. S. PELLINI, "Soundness of Cast Steel Bars," *Trans. A.F.S.* **59**, 174 (1951).

4. H. F. BISHOP and W. S. PELLINI, "The Contribution of Riser and Chill Edge Effects to Soundness of Cast Steel Plates," *Trans. A.F.S.* **58**, 185–197 (1950).

5. *Steel Casting Handbook*, Steel Founders Society of America, Cleveland, O., 1950.

6. R. P. DUNPHY and W. S. PELLINI, "A Solidification Dilatometer and Its Application to Gray Iron," *Trans. A.F.S.* **60**, 783–788 (1952).

7. W. A. SCHMIDT, E. SULLIVAN, and H. F. TAYLOR, "Risering of Gray Iron Castings," *Trans. A.F.S.* **62**, 76 (1954).

8. W. S. PELLINI, "Factors Which Determine Riser Adequacy and Feeding Range," *Trans. A.F.S.* **61**, 61–80 (1953).

9. H. F. BISHOP and W. H. JOHNSON, *The Foundry*, Vol. 84, No. 2, p. 70 (1956); No. 3, p. 136 (1956).

10. R. A. FLINN and C. F. MIELKE, "Pressure Tightness of 85-5-5-5 Bronze, Thermal Gradients During Solidification," *Trans. A.F.S.* **67**, 385–392 (1959).

GENERAL REFERENCES

Transactions A.F.S., The Foundry, and R. W. RUDDLE, "Risering of Castings," in *The Running and Gating of Castings*. The Institute of Metals Monograph and Report Series **19** (1956).

1. Using (a) the Caine method, (b) the NRL method, design a risering system, taking into account ease of riser removal and molding procedure, that will produce sound castings with maximum yield for the following simple shapes:

    (i) a plate of dimensions $10 \times 10 \times 20$ in.,

    (ii) a cube of dimensions $6 \times 6 \times 6$ in.,

    (iii) a ball-mill liner of dimensions $50 \times 60 \times 2$ in.,

    (iv) a pipe 30 in. long; outside diameter 5 in., inside diameter 3 in.

2. (a) Calculate an economical risering system that will ensure soundness in the casting (0.30% steel) shown in Fig. 3–16.

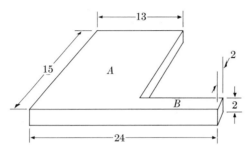

FIGURE 3–16

(b) Calculate an economical risering system for the same casting, assuming that the density increase from liquid to solid is very slight, but not negligible (0.1%), that there is no dendritic growth during solidification, and that the metal is pure.

CHAPTER 4

GATING DESIGN

4-1 Introduction. The objective of a gating system is to permit distribution of the metal to the mold cavity at the proper rate, without excessive temperature loss, free from objectionable turbulence, entrapped gases, slag, and dross. The problem at hand, therefore, is to study the available theoretical and experimental data so that we can make an intelligent choice of gating-system design for a given combination of metal, mold material, and molding process.

It must be recognized at the outset that any good gating system is the result of considered engineering compromises. Earlier, we mentioned briefly that metal and mold compositions affect the choice of design for a gating system. For example, we can and should employ a more elaborate design to avoid dross (oxides, etc.) in an easily oxidized metal of low melting point, such as aluminum, in contrast to the short, more turbulent system that is suitable for a nondrossing alloy of high melting point, such as cast iron, for which it is advantageous to avoid the higher pouring temperatures required by a long metal path. Similarly, the characteristics of a heated ceramic mold such as is used in investment casting permit variations in the gating system customarily used with cold molds of other compositions.

Before proceeding further the reader may wish to review the specialized nomenclature of gating (Appendix II) as standardized by the American Foundrymen's Society.

The basic starting point for any discussion of gating is a review of the principles of fluid flow for vertical and horizontal passages. These calculations are useful for estimating the pouring time of a casting and, what is far more important, by giving advance warning of undesirable turbulence and aspiration of mold gases, they enable the engineer to prevent the development of these conditions. Our emphasis will be quite different from that in the study of hydraulics, where interest is centered on the accurate metering of fluid flow. It makes little difference to us whether or not a pouring system is designed to yield an accurate pouring time, but it is vitally important to minimize aspiration of mold gases and to control the metal distribution in the mold.

We shall first discuss the basic equations of flow and then apply them to the vertical and horizontal portions of the gating system. Finally, we shall consider some special cases in actual mold design that involve step, multiple, and riser gates, and the effects of mold tilting. From elementary

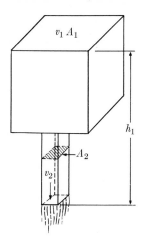

FIG. 4–1. Area-velocity relations in simple impermeable mold.

hydraulics, we can apply two very important principles, *the law of continuity* and *Bernoulli's equation*.

4–2 Law of continuity. The law of continuity states that for a system (like that shown in Fig. 4–1) with *impermeable* walls and filled with an incompressible fluid,

$$Q = A_1 v_1 = A_2 v_2, \tag{4–1}$$

where Q = rate of flow, A = area, and v = velocity. Since liquid metals are only slightly compressible, they may be treated as incompressible fluids for present purposes. In dealing with the casting of metals, it is usually convenient to use the inch as the unit of distance and the second as the unit of time. Therefore Q is expressed in cubic inches per second, A in square inches, and v in inches per second. Thus, given that Fig. 4–1 represents a full liquid column, and that $A_2 = 0.5 \text{ in}^2$, $A_1 = 2 \text{ in}^2$, and $v_1 = 10 \text{ in/sec}$, it follows that $v_2 = 40 \text{ in/sec}$.

Note that although there may be turbulence and frictional losses, Eq. (4–1) nonetheless holds for an incompressible fluid so long as the system is full and the walls are impermeable. However, because sand molds are permeable, there are several important exceptions which we shall discuss shortly.

4–3 Bernoulli's equation. This expression, which relates the pressure, velocity, and elevation at points along a line of flow, is easily derived by applying the principle of the conservation of energy. In the system shown in Fig. 4–2 the energy of the stream at position 2 is equal to the energy at position 1, less any energy losses due to turbulence which occurred in passing from 1 to 2:

$$E_1 = E_2 + E_{\text{lost}(1-2)} \qquad (4\text{-}2)$$

For purposes of calculation, the energy can be represented more conveniently by the "head" or distance above a datum plane, usually selected at the point of efflux (such as the ingate to a casting). If a particle of the fluid possesses kinetic energy, this is converted by calculation to equivalent potential energy. Thus a stationary particle of unit weight at position 1 would have *potential energy* which would be related to the height h above the datum plane. If frictional losses are neglected, a similar particle at 2 would have *kinetic energy* equal to the potential energy at 1. Also, a stationary particle at 3 would have *pressure energy* equal to the potential energy at 1 or the kinetic energy at 2. The pressure energy is transformed to an equal amount of kinetic energy as the particle passes into free space.

Therefore, if no frictional losses are involved, $E_1 = E_2 = E_3$ or, in general, the sum of the potential-energy, the pressure-energy, and the kinetic-energy terms for any point in the same system is equal to that for any other point. For example, the energy at point 1 equals the energy at point 2, and therefore Bernoulli's equation can be written as

$$h_1 + \frac{v_1^2}{2g} + \frac{p_1}{w} = h_2 + \frac{v_2^2}{2g} + \frac{p_2}{w}, \qquad (4\text{-}3)$$

where

$h = $ the potential energy, in ft or in,

$\dfrac{v^2}{2g} = $ the kinetic energy, in ft or in,

$v = $ velocity, in ft/sec or in/sec,

$g = $ the acceleration due to gravity $= 32.2 \text{ ft/sec}^2$

$\quad = 386 \text{ in/sec}^2$,

$\dfrac{p}{w} = $ the pressure energy, in ft or in,

$p = $ pressure, in lb/ft^2 or lb/in^2,

$w = $ density of fluid, in lb/ft^3 or lb/in^3,

atmospheric pressure $= 14.7 \text{ lb/in}^2$,

density of liquid iron $= 0.26 \text{ lb/in}^3$.

The potential energy h is usually taken as zero at the point of efflux, but p is always taken as the actual pressure. For example, at 1 atm, $p = 14.7 \text{ lb/in}^2 = 2120 \text{ lb/ft}^2$.

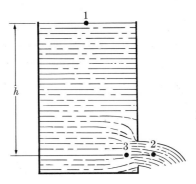

FIG. 4–2. Interrelation of potential energy and kinetic energy in flow.

For the tank shown in Fig. 4–2, assuming no frictional losses, we calculate energies at points 1 and 2 as follows:

$$h_1 + 0 + \frac{p_1}{w_1} = 0 + \frac{v_2^2}{2g} + \frac{p_2}{w_2},$$

and since $p_1 = p_2$ and $w_1 = w_2$,

$$v_2 = \sqrt{2gh_1}. \qquad (4\text{--}4)$$

Now let us apply these principles first to vertical and then to horizontal gating systems.

VERTICAL GATING

4–4 Example of vertical gating. When h and v are given in inches and inches/second respectively, the velocity of a metal stream in a pressurized (i.e., filled) gating system is

$$v = 27.7\sqrt{h} \text{ in/sec.} \qquad (4\text{--}5)$$

As an example, let us calculate the pouring time of a $5 \times 10 \times 20$ in. casting (Fig. 4–3) using 1×1 in. ingate (or opening to the mold cavity) and a constant head of 5 in. (neglect orifice and frictional effects). We obtain

$$t = \frac{V}{Q} = \frac{V}{Av} \qquad (V = \text{volume})$$

or

$$\text{total time} = \frac{\text{total volume}}{\text{in}^3/\text{sec}} = \frac{1000 \text{ in}^3}{1 \text{ in}^2 \times 27.7\sqrt{5} \text{ in/sec}}$$

$$= 16.1 \text{ sec.}$$

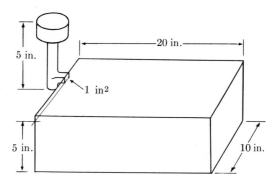

FIG. 4–3. Illustration of a simple gating calculation.

This example is merely for purposes of illustration. Actually such a system would be a poor design, as we shall explain in Section 4–7.

4–5 Aspiration effects. So far we have considered only the flow of metal within impermeable walls. Since most castings are poured in permeable sand molds, we now must examine the effect of gate design upon the aspiration of gases from the mold into the metal. These gases can be air, water vapor, or decomposition products of organic materials used to bond the mold. Let us first consider the case of the straight downsprue (Fig. 4–4) and begin with an impermeable mold, not with sand walls as shown. For simplicity, we shall assume that the metal height, h_c, in the pouring cup is kept constant. Let us first examine the pressure head at different parts of the system, under the assumption of *impermeable* walls and with the top (1) and the bottom (3) of the system at atmospheric pressure.

For cast iron ($w = 0.26 \text{ lb/in}^3$), Bernoulli's equation for points 1 and 3 is

$$ \overset{(1)}{h_t + 0 + \frac{14.7}{0.26}} = \overset{(3)}{0 + \frac{v_3^2}{2g} + \frac{14.7}{0.26}}, $$

or

$$ v_3 = 27.7\sqrt{h_t} \text{ in/sec} \qquad \text{(friction neglected)}. $$

Thus the stream issues from point 3 in Fig. 4–4 at $27.7\sqrt{h_t}$ in/sec and at atmospheric pressure. However, by the law of continuity, the velocity at point 2 must also be $27.7\sqrt{h_t}$ in/sec. This seems at first to disprove the principle of conservation of energy, since point 2 is at a higher level than point 3 and so has greater potential energy. The inequality arises

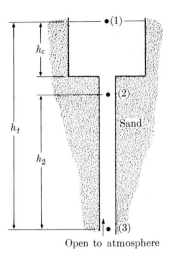

Open to atmosphere

FIG. 4–4. Simple vertical gating system in permeable mold.

from the pressure term:

$$\underset{(2)}{h_2 + \frac{v_2^2}{2g} + \frac{p_2}{w}} = \underset{(3)}{0 + \frac{v_3^2}{2g} + \frac{14.7}{w}} ,$$

where w = density of metal, in lb/in^3. Then, since $v_2 = v_3$,

$$p_2 = 14.7 - h_2 w. \tag{4–6}$$

Therefore the pressure throughout the liquid at point 2 is *less than at-mospheric* by the factor $h_2 w$. Note that this calculation is for an impermeable mold. Now the air or mold gas at the sand-metal interface in a permeable mold is at least at atmospheric pressure (pressure is usually above 1 atm, since mold gases are generated by the action of the hot metal and even the air pressure is increased). Gas will, therefore, be aspirated into the metal stream at 2. The quantity of gas will depend upon $h_2 w$, the permeability of the mold, and the pressure of the mold gas. The fate of the mold gas is varied: it may react with the metal, forming oxides and dross, dissolve in the metal to precipitate later upon freezing, or remain in the metal in the form of mechanically entrapped bubbles.

All these possibilities are undesirable, but can be better tolerated in some metals than in others. In cast iron, the oxygen of a bubble of air may partly react with the metal to form fine harmless silicate particles, and the nitrogen may either dissolve or bubble out if the casting solidifies slowly. In aluminum, however, an oxide film of about the same density as that of aluminum may be trapped and thus contain the nitrogen bubble. The high surface tension of the aluminum-oxide surface will retard the escape of the bubble and a hole may result. In addition, hydrogen may be dissolved by decomposition of the water vapor coming from the mold or the air.

4–6 Prevention of aspiration. (a) To determine whether the downsprue of a permeable mold can be designed to prevent aspiration, let us reexamine the Bernoulli equation for points 2 and 3 in Fig. 4–4:

$$h_2 + \frac{v_2^2}{2g} + \frac{p_2}{w} = 0 + \frac{v_3^2}{2g} + \frac{p_3}{w}.$$

We see that if p_2 is made equal to p_3 and equal to 1 atm to avoid aspiration, we obtain

$$\frac{v_3^2}{2g} = h_2 + \frac{v_2^2}{2g}.$$

Also,

$$A_2 v_2 = A_3 v_3,$$

$$v_2 = \frac{A_3}{A_2} v_3.$$

If we denote A_3/A_2 by R, then $v_2 = Rv_3$ and

$$R^2 = 1 - \frac{2gh_2}{v_3^2}.$$

Now, since $h_2 = h_t - h_c$ and $v_3 = \sqrt{2gh_t}$,

$$R = \sqrt{h_c/h_t}. \tag{4–7}$$

For example, in Fig. 4–4, let $h_2 = 6$ in., $h_t = 8$ in. and $h_c = 2$. We wish to calculate the ratio of the areas of the downsprue at points 2 and 3, to prevent aspiration at 2. Applying Eq. (3–7), we find

$$\frac{A_3}{A_2} = R = \sqrt{2/8} = \frac{1}{2}.$$

Hence, for this case, A_2 should be twice A_3.

(b) The calculation of the area ratios can also be approached from another point of view which may give a better physical conception of the problem. Let us consider the walls to be completely permeable and examine the shape of a freely falling column issuing from the pouring cup, as shown in Fig. 4–5. Then

$$v_2 = \sqrt{2gh_c},$$

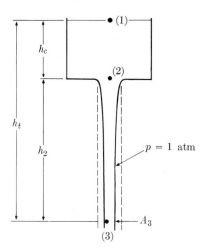

FIG. 4–5. Stream shape in open mold.

where h_c is the height of the metal in the pouring cup, and

$$v_3 = \sqrt{2gh_t}.$$

But since $Q_2 = Q_3$, we have

$$A_2 v_2 = A_3 v_3,$$

where A is taken as the cross section of the *stream,* not of the downsprue. Therefore

$$\frac{A_3}{A_2} = \frac{V_2}{V_3} = \sqrt{\frac{h_c}{h_t}}. \tag{4–8}$$

The same ratio is obtained by both methods because, if the permeable mold wall in case (b) is molded to fit the contour of the stream, p will continue to be one atmosphere throughout and no aspiration will occur. It should be noted that the solution of the area relationship for locations lying between points (2) and (3) describes a hyperbolic section. However, in practice, A_2 and A_3 may be joined by straight lines, since one thus obtains an even greater area in the upper regions, resulting in lower velocities and higher pressures. This also leads to simpler pattern construction.

An important corollary is that if a *straight* or reverse-taper downsprue is used in a permeable mold, the rate of flow is governed by the *head in the pouring cup only.* This problem is essentially that of Fig. 4–2, where the orifice is placed at the bottom of the tank. The authors of some publications in the foundry field have been mystified by the fact that an increase in h_2 does not increase the flow rate from a straight-walled downsprue. The reason should now be obvious.

One other effect should be mentioned briefly. The quantity of metal Q issuing from the bottom of the pouring cup is controlled entirely by h_c in a straight or reverse sprue system, when the metal in the cup is quiet. However, if the stream of metal from the ladle impinges at the top of the sprue, the kinetic energy of the stream is converted partially to pressure energy at this point, and this effect increases the velocity at 2. The increment is difficult to estimate, and such pouring practice is usually not recommended.

4–7 Bottom gating systems. So far, we have considered vertical gating systems with atmospheric pressure only at the base. In many castings it is advantageous to place the ingate at the bottom of the mold cavity. In this way the splashing and oxidation accompanying top gating are avoided.

The calculation of pouring time must now be modified. Let us consider a nonaspirating downsprue operating with head h_t to fill the mold cavity, as in Fig. 4–6. As soon as the mold begins to fill, the effective head decreases. Assuming a mold of simple shape with parallel sides and no frictional losses, let

$$t = \text{time elapsed from start of pouring,}$$
$$h_t = \text{total head,}$$
$$h = \text{height of metal in mold,}$$
$$A_m = \text{cross-sectional area of } mold \text{ (horizontal)}$$
$$\text{(i.e., horizontal cross section of casting),}$$
$$A_g = \text{cross-sectional area of gate.}$$

In an increment of time dt, the height will increase dh, and the volume of metal will increase $A_m \, dh$. Also, the amount delivered by the gate in time dt will be $A_g v \, dt$, where A_g and v are the area and instantaneous velocity at the gate, respectively. The velocity at the gate is given by the

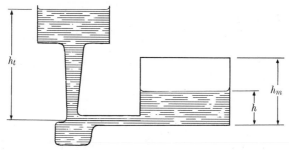

Fig. 4–6. Nonaspirating downsprue for permeable mold.

expression $\sqrt{2g(h_t - h)}$. Then, equating the increase in casting volume in time dt to the flow through the ingate in the same interval, we obtain

$$A_m \, dh = A_g \sqrt{2g(h_t - h)} \, dt \qquad \text{and} \qquad \frac{dh}{\sqrt{2g(h_t - h)}} = \frac{A_g}{A_m} \, dt.$$

If t_f is the time required to fill the mold cavity and h_m is the height of the mold cavity, we can integrate between $t = 0$ and $t = t_f$ and between $h = 0$ and $h = h_m$:

$$\frac{1}{\sqrt{2g}} \int_0^{h_m} \frac{dh}{\sqrt{h_t - h}} = \frac{A_g}{A_m} \int_0^{t_f} dt,$$

$$t_f = \frac{2A_m}{A_g \sqrt{2g}} \left(\sqrt{h_t} - \sqrt{h_t - h_m} \right). \tag{4–9}$$

It is evident that the increase in pouring time due to bottom gating is significant. If in Fig. 4–3 the entire sprue were moved to a bottom gating position, we would have

$$h_t = h_m = 5 \text{ in,}$$

and thus

$$A_m = 200 \text{ in}^2, \qquad t_f = \frac{(2)(200)}{1\sqrt{2g \cdot 12}} (\sqrt{5} - 0) = 32.2 \text{ sec.}$$

This is twice the pouring time required for the top gating. It is left as an exercise to prove that this is always true for $h_t = h_m$.

In many molds a modification of the above situation occurs if open risers are used, since these are filled to the level of the pouring sprue. In this case, one first calculates the pouring time of the casting and then the filling time of the riser, using the cross-sectional area of the riser instead of A_m. The sum of these times gives the total pouring time.

As h_t increases relative to h_m, the expression $(\sqrt{h_t} - \sqrt{h_t - h_m})$ decreases, leading to shorter pouring times. This is not a linear relationship; it can be shown by graphical methods that when t reaches $t_{max}/2$ ($t = t_{max}$ at $h_t = h_m$), there is relatively little decrease in time with further increase in h_t.

HORIZONTAL GATING

4–8 Function of the horizontal gating system. The purposes of the horizontal portion of the gating system are:

(1) To introduce the metal to the mold cavity in the best distribution pattern, with minimum turbulence and temperature loss.

(2) To reduce and trap impurities in the metal stream rather than increase them.

Some castings are poured without horizontal gates; the metal falls directly into the casting cavity from pencil gates in the cope surface. This can only be done with metals which oxidize slowly and are not susceptible to gas pickup. The effect of the falling metal upon the drag surface is not so severe as expected because the metal to enter first provides a cushion on the floor of the cavity for the remaining metal. The advantages of this practice are reduced molding cost and higher yield. The car wheel illustrated in Fig. 1–5 is a case of this type.

The great majority of castings, however, are poured with horizontal gating, and some of the principles involved therefore deserve attention here.

4–9 Aspiration effects at points of change in metal stream direction.

When a metal stream passes through an orifice or around a sharp corner, momentum effects result in a contraction of the stream as shown in Fig. 4–7(a). The constricted region is known as the *vena contracta*.

If we write Bernoulli's equation for points 2 and 3 (Fig. 4–7a), we obtain

$$0 + \frac{p_2}{w} + \frac{v_2^2}{2g} = 0 + \frac{p_3}{w} + \frac{v_3^2}{2g}.$$

Since $p_3 = 1$ atm and since, for continuity, the velocity at 2 must be greater than at 3, we may rewrite the above relationship as follows:

$$\frac{p_2 - p_3}{w} = \frac{v_3^2 - v_2^2}{2g}.$$

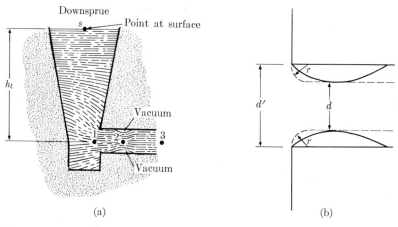

Fig. 4–7. (a) Orifice effects in impermeable mold. (b) Runner (dashed lines) designed to prevent aspiration.

Then, solving for p_2, we get

$$p_2 = p_3 + w\left(\frac{v_3^2 - v_2^2}{2g}\right). \tag{4–10}$$

Since $v_3 < v_2$, we have $p_2 < p_3$ (1 atm). In other words, p_2 will be reduced below atmospheric pressure, as a function of the stream velocity and the degree of stream contraction.

This case is quite different from that of vertical gating. The reason for the vena contracta is not acceleration due to gravity but simply the change in flow direction. The remedy is also different. Instead of a sharply edged orifice, the flow lines should be curved to cause a gradual change in direction. (Note that a mold made to fit the vena contracta would not raise the pressure at 2.) However, once this gentle change in direction has been accomplished, there is no need to enlarge the section again. A suggested design, shown in Fig. 4–7(b), is based on the following reasoning. Assume that to provide the desired flow rate, we wish to have a runner of diameter d. To avoid the reduced pressure, we shall design the entrance to the runner with a radius r which is even greater than called for by the natural vena contracta. To calculate r we use the empirical observation that the highest ratio for d'/d encountered in gating systems is of the order of 1.30. Since $d' = 1.30d$, $r = 0.15d$. Therefore for a runner 2 in. in diameter, $r = 0.3$ in.

Changes in direction are, in general, encountered at two locations: at the junction between vertical and horizontal gates and at the mold cavity. It is simple and inexpensive to provide proper junctions between gating systems. On the other hand, the use of a flared ingate adds considerably to the cleaning expense. This disadvantage can be remedied by employing a slow-moving stream as discussed in the next section.

Actual Gating Systems

Although there seems to be an infinite variety of gating systems, there are really only four basic kinds: (1) flat, platelike castings molded with the largest dimension in the horizontal plane, gated at the parting line; (2) complex chunky castings, often with many cores or cheek flasks, bottom-, top-, or side-gated; (3) stack molds, which are combinations of (1) and (2); (4) dross trap systems.

4–10 Platelike castings. The greatest amount of experimental work has been done with castings of this type, particularly with lucite molds poured with water [2]. A typical recommended system is illustrated in Fig. 4–8. Three features of this system merit special attention.

(a) Tapered downsprue. For straight downsprues, experimental data show severe aspiration, in confirmation of the Bernoulli analysis.

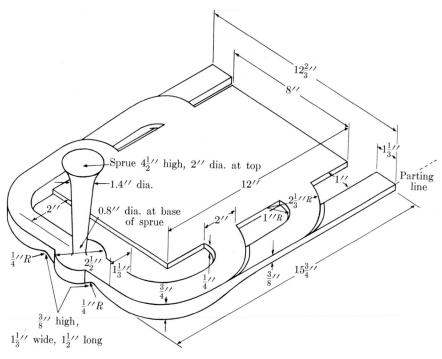

FIG. 4–8. Preferred gating system for platelike castings [2].

(b) Well at downsprue base. After testing many different designs of downsprues, the type shown in Fig. 4–9 was developed. The general requirements are that the extension of the sprue beneath the runner should be greater than the runner depth, and the horizontal cross-sectional area of the well should be more than five times the sprue area at the base.

(c) Sprue/runner/gate area ratios. The most satisfactory ratio of sprue area to runner area to gate area was found to be 1 : 4 : 4, where the sprue area is calculated at the sprue base. Since the pressure in the stream at the ingate to the casting cavity is atmospheric, higher upstream pressures are required to make up for pressure losses at bends, sidewalls, etc. The runner is "pressurized" because its section is reduced more drastically than warranted by the decrease in the volume of metal flowing through the gates.

4–11 Complex chunky castings. In a complex casting, it is desirable to deliver metal to each particular level or protuberance, to avoid misruns or cold shuts. The principal method used has been step gating. Many combinations of this type of gate (Fig. 4–10) have been investigated [3]. These entirely unsuccessful experiments show clearly that the

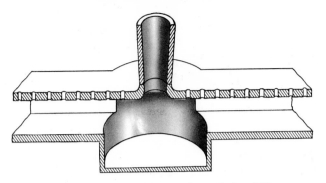

FIG. 4–9. Preferred well-base design [2].

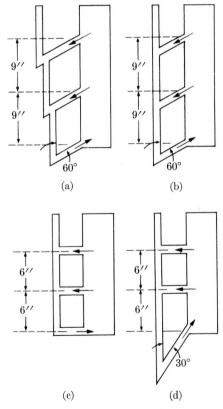

FIG. 4–10. Step-gating systems [3].

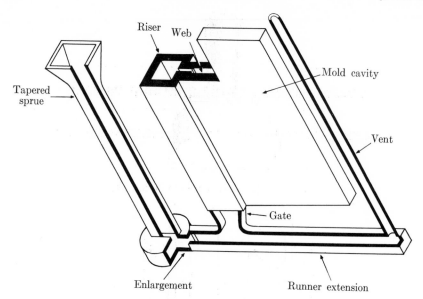

FIG. 4–11. Upsprue with web gate.

metal does not follow the wishes of the designer, but rather obeys the laws of momentum. Several kinds of combination runners were tried without success. Only when the runners were connected directly to an upsprue, rather than to the downsprue, was satisfactory performance obtained. A comparable situation, in which the upsprue, or web, is connected continuously to the casting with a thin gate [2], is illustrated in Fig. 4–11.

4–12 Stack molds. The stack-molding method is economical because it is possible to produce one complete mold per flask section. The bottom of a given flask in a stack provides the cope of the flask below, as shown in Fig. 4–12, and the top provides the drag of the next layer. In the usual stack mold the castings are gated from runners radiating from a single long downsprue.

With a drossy metal or when progressive solidification is required, the turbulence and uneven metal delivery to the side runners from the downsprue are quite serious. As shown in our earlier analysis, the rate of filling is not proportional to the total sprue height, but only to the height of metal in the pouring cup. Due to momentum effects, the metal will pulsate in the downsprue and splash into several mold layers at once.

To avoid gating directly from the downsprue, the system shown in Fig. 4–12 has been developed. The large downsprue leads to four runners in the base of the mold, which in turn fill four upsprues. The runner area

FIG. 4–12. Stack molding of automobile tappet bases (approx. ⅓ size).

is calculated to provide a choke at the base of the downsprue. Therefore the metal flows smoothly in the downsprue and rises evenly in the up-sprue, and the castings are poured in smooth regular fashion, beginning with the bottom layer and progressing to the top. (It is not possible to provide a tapered downsprue, since each flask section, except the special bottom plate, is made from the same pattern. This is why the choked downsprue is employed.)

4–13 Dross trap system. Thus far we have considered only the flow after a system has been filled. Regardless of the type of system, the first metal stream to enter will be broken up by the momentum effects at corners and will probably contain dross and slag. Momentum effects can also carry dross past the gates into a blind alley, where it will remain. The metal following it will be of better quality if the appropriate gating is used (Fig. 4–8). Note that in this system which is the preferred gating for a flat plate, the runners are in the drag, and the gates in the cope. This is done to prevent the first metal passing through the runner from

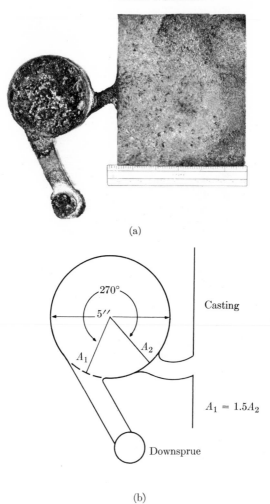

(a)

(b)

Fig. 4–13. (a) Whirl gate for entrapment of slag and dirt. To simulate severe slag conditions, glass beads were added to the stream of metal entering the downsprue. Note that these have collected as a glassy mass in the center of the cylinder and that more are present in the cope surface of the casting. (b) Design details of whirl gate.

entering the mold. The advantages of this design are the relatively low velocity at which metal enters the mold, and the lack of aspiration in the gating system.

Another trap system which has proved quite successful in removing slag from steel is the 270° whirl gate (Fig. 4–13) which utilizes the principle of centrifugal action to whirl the less dense materials to the center of the cylinder. To accomplish this effect the following details are important. The area A_1 of the ingate to the whirl should be at least 1.5 times the exit gate, A_2, so that metal builds up in the cylindrical portion. The entering metal should revolve 270° before reaching the exit gate, to provide time for whirling impurities to the center. A slight offset in the cylinder wall next to the exit is desirable to drive the first metal entering the whirl past the exit for further centrifugal action.

REFERENCES

1. R. W. RUDDLE, "The Running and Gating of Castings," The Institute of Metals, Monograph and Report Series No. 19 (1956).

2. W. L. EASTWOOD, "Tentative Design of Horizontal Gating Systems for Light Alloys," Symposium on Principles of Gating, A.F.S. (1951).

3. W. H. JOHNSON and W. O. BAKER, "Gating Systems for Metal Castings," *Trans. A.F.S.* **56**, 389–397 (1948).

GENERAL REFERENCES

Transactions A.F.S., The Foundry, and R. W. RUDDLE, *The Solidification of Castings.* London: The Institute of Metals, 1956.

PROBLEMS

1. For the castings of 0.3% carbon cast steel shown in Figs. 4–14 and 4–15 calculate and draw risering and gating systems to obtain radiographic soundness in all sections with maximum casting yield. A pouring time of 3 sec is desired in each case. Calculate yield as follows:

$$\frac{\text{weight of casting}}{\text{total weight poured in mold}}.$$

Consider in each case how the risering might be changed if radiographic quality were not essential. What changes would you make for a high-carbon cast iron?

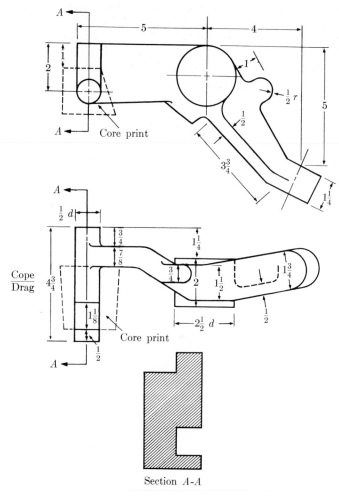

Section *A-A*

FIG. 4–14. Lever casting.

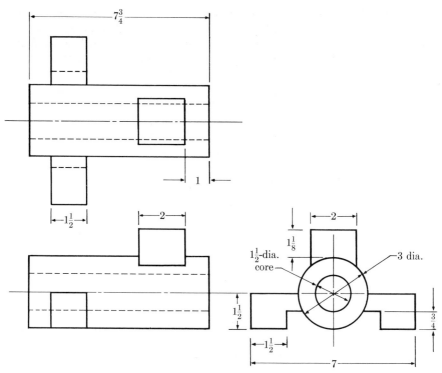

Fig. 4–15. Bearing casting.

2. Calculate A_2 required to pour the entire mold (riser and casting) in 10 sec without aspiration after the downsprue has been filled. Assume that the ingate is at the cope surface of the casting. Neglect *frictional* and *orifice* effects.

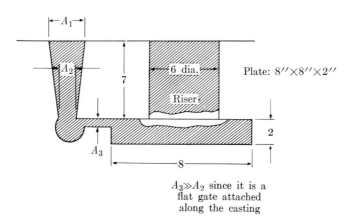

Plate: $8'' \times 8'' \times 2''$

$A_3 \gg A_2$ since it is a flat gate attached along the casting

Figure 4–16

3. Neglecting wall friction and orifice effects, calculate the proper design of the downsprue shown to deliver liquid cast iron of density 0.25 lb/in^3 at a rate of 20 lb/sec against no head at the base of the sprue. The height of the metal in the sprue cup and the other dimensions listed are fixed by the flask in use.

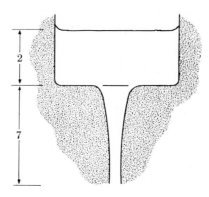

FIGURE 4-17

4. In the pouring cup-downsprue system shown, derive by two methods the relationship between A_1 and A_2 which will avoid aspiration: (a) Use the Bernoulli equation; (b) calculate the mold contour to fit the falling stream (neglect orifice and wall friction effects).

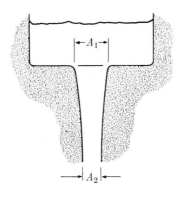

FIGURE 4-18

CHAPTER 5

METAL FLUIDITY

5–1 General. Let us assume that we have completed a mold design based on the preceding calculations of risering and gating. The major problem remaining is the selection of the proper pouring temperature for the alloy specified. To understand this problem, we must recall that in our discussion of gating we dealt with liquid metal as a typical fluid. However, there are two important qualifications to be added: (a) If an alloy is poured at too low a temperature, it will begin to solidify in the gating system and the mold may not be filled. A defective casting of this type is called a *misrun*. Other defects caused by low pouring temperatures are laps and seams. In these cases the casting contains mechanical faults caused by incomplete bonding between two different streams of liquid metal flowing into the mold cavity. (b) If a metal is poured at an excessively high temperature, the sand in the gating system and in the mold walls will be attacked more strongly. This can result in rough surfaces on the casting and in entrapment of the reaction products, solids, liquids and gases, in the casting.

It is evident therefore that a pouring temperature should be selected which will avoid both types of defects. Furthermore, because of the usual production conditions under which many castings are poured from a single ladle, it is necessary to specify a pouring-temperature range. The width of this range should not be left to the discretion of the pourer, who will always ask for the widest possible range, but should be decided upon after actual determination of the range encountered during proper pouring from a well-preheated ladle. The narrower the range that can be maintained, the more consistent will be the surface finish, soundness, and performance of the resulting castings.

5–2 Measurement of fluidity. The term *fluidity* as used in the cast-metals industry is quite different in meaning from the term defined in physical chemistry. In the latter case fluidity is the reciprocal of viscosity. By contrast, the foundryman is in search of a quantity that will permit him to evaluate the relative ability of a given liquid alloy at a certain temperature to fill a mold. In the following discussion it will be seen that the foundry definition must take into consideration the effect of decreasing temperature which tends to produce crystallization in the metal stream and thus impede its flow.

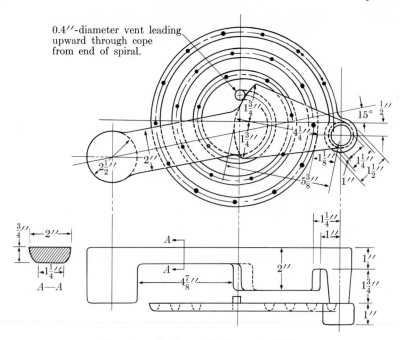

0.4″-diameter vent leading upward through cope from end of spiral.

FIG. 5–1. Design of fluidity spiral [2].

In other words, the viscosity of any liquid metal is low enough (i.e. the fluidity as conventionally defined is high enough) at any temperature above the liquidus to permit successful filling of the mold [1]. Unfortunately, therefore, there is no simple concept which will express the "fluidity" as the term is used in the foundry and in the sense in which it will be used throughout this discussion.

To provide a relative measure of fluidity a variety of empirical methods have been developed. The two which have recently received most attention are the sand-cast fluidity spiral and the suction-tube method. We shall discuss the data obtained with each type separately and then attempt to reconcile the differences by means of solidification theory.

5–3 The sand-mold fluidity spiral. Since the major difficulties usually arise in connection with the thin sections of a casting, the earliest designs of fluidity-test castings were simply long, thin straight sections. Finally, over the past thirty years, a more compact and reproducible design has been developed (Fig. 5–1), consisting of a spiral 55 inches long which can be filled completely only under conditions of very high fluidity. The length of the spiral obtained, in the usual case, from the partially filled mold, provides an index of relative fluidity. In the original design [6]

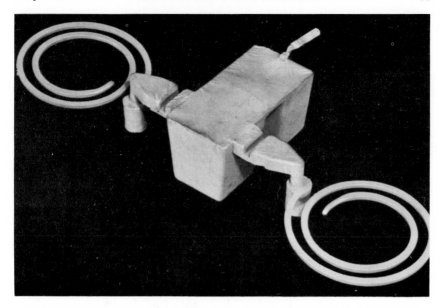

Fig. 5–2. Fluidity spiral [3].

reproducibility varied widely because the effective pouring head depended upon the speed of the operator in filling the pouring basin. Secondly, most investigators used a thermocouple with a heavy protection tube which resulted in a severe temperature lag and improper temperature readings. The procedure developed by Mott, Schaefer, and Cook [3] eliminated both of these variables for the first time. The mold design shown in Fig. 5–2 is based on their technique for duplicate spirals. The mold is poured under a head, which is kept constant by virtue of the mold design. In previous procedures the operator could splash the first metal from the ladle into the mold and knew when it was being filled. In the design shown, the operator pours for several seconds before the orifice to the downsprue in the side of the basin is reached. As soon as the orifice is filled, the resulting head cannot be exceeded by more than $\frac{1}{4}$ inch because any further excess would flow off into the overflow basin. The pourer does not know exactly when the mold is being filled, and the pouring becomes objective and mechanical.

Temperature measurement is accomplished by means of a fine-wire platinum-10% rhodium-platinum thermocouple enclosed in a thin-walled fused silica tube. The thermal lag of this couple is approximately one second, whereas the heavier types formerly used exhibited a lag of the order of 30 seconds. After an accurate correlation of optical-pyrometer and thermocouple measurements has been obtained, the optical pyrometer

TABLE 5–1

FLUIDITY VS. LIQUIDUS TEMPERATURE

Iron	Pouring temp. for 30-in. spiral,		Difference above Iron 1,		Liquidus		Liquidus, degrees above Iron 1,	
	°F	°C	°F	°C	°F	°C	°F	°C
(1) 3.6%C, 2.08%Si	2370	1299	—	—	2150	1177	—	—
(2) 3.04%C, 2.10%Si	2480	1342	110	43	2270	1226	120	49
(3) 2.52%C, 2.00%Si	2610	1415	240	116	2360	1276	210	99
(4) 2.13%C, 2.07%Si	2660	1444	290	143	2425	1312	275	135

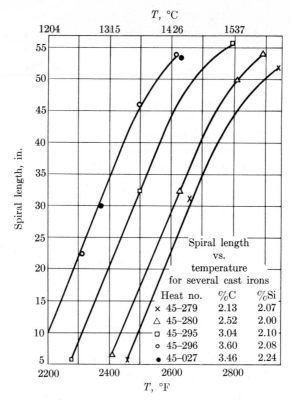

FIG. 5–3. Typical temperature-fluidity curves [2].

may be employed as the temperature indicator for certain alloys. However, for alloys forming an oxide film which prevents sighting of the clean metal surface, thermocouple measurements should be used exclusively.

Another important characteristic whose significance will become clear in later discussions, is the freezing point or start of the freezing range (liquidus). This information can be obtained by recording the thermocouple readings versus time. If the thermocouple technique becomes awkward or expensive, an approximate, visual method has been devised which is quite satisfactory. An open mold, approximately three inches in diameter and two inches high, is poured at a temperature at least 150°F (83°C) above the liquidus, and the metal temperature is read continuously with an optical pyrometer. The top surface is continually cleared with a light rod or wire until the cooling is finally arrested. This stage coincides with the formation of small amounts of solid at the surface which can be observed independently as a check on the arrest temperature. The technique described can be quite accurate, because the average arrest lasts about ten seconds to one minute and affords sufficient time for determining the temperature.

5–4 Typical fluidity curves. Typical data for the actual relationship of fluidity to temperature and composition are shown in Fig. 5–3. A series of spirals poured from the same ladle of metal but at different temperatures approximates a linear variation of spiral length with temperature, particularly at the shorter spiral lengths. By extrapolation one can show that the lines in Fig. 5–3 reach zero fluidity at close to the freezing point of the particular alloy. The curves also indicate that practically any desired degree of fluidity can be obtained in any alloy, provided that the proper pouring temperature is attained. Therefore, although a low-carbon cast iron, or steel for that matter, is usually rated as a low-fluidity material, it is only the limitations of both furnace equipment and refractoriness of molding sands that lead to this classification.

5–5 Effect of metal chemistry. The fact that the fluidity graphs shown in Fig. 5–3 are parallel leads us to seek a general relationship of fluidity versus temperature and composition. If a constant fluidity line is drawn across the graph, at 30 inches for example, the various irons show the pouring temperatures and freezing-point relationships listed in Table 5–1. The increase in pouring temperature required to obtain the same fluidity in Iron 4 as in Iron 1, for example, is approximately equal to the difference in the freezing points (290°F, 161°C) of the two metals. In other words, fluidity as defined by the foundryman is the summation of

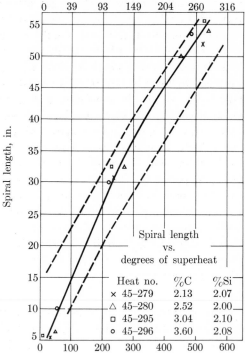

FIG. 5–4. Fluidity versus °F superheat above liquidus [3].

the fluidities existing over the entire temperature range, from the pouring temperature to solidification. Since only minor changes in viscosity take place until the freezing range is reached, the castability depends principally on the superheat, in °F, of the pouring temperature above the freezing range. If an appreciable change in viscosity, or the conventional "fluidity," had occurred with decreasing temperature, the integrated change in fluidity over the whole temperature range would have determined the length of the fluidity spiral.

Inasmuch as fluidity is a function of the superheat above the liquidus, fluidity data from all the irons can be plotted on a common basis as shown in Fig. 5–4, and the result is one curve. The question now arises as to the applicability of this curve to materials other than cast iron. In the same figure, points are plotted for alloy steels and silicon copper, and only a relatively narrow scatter band is obtained. Experiments recently completed show that even metals with low melting points, such as aluminum and lead, follow this curve [3]. One characteristic deviation has

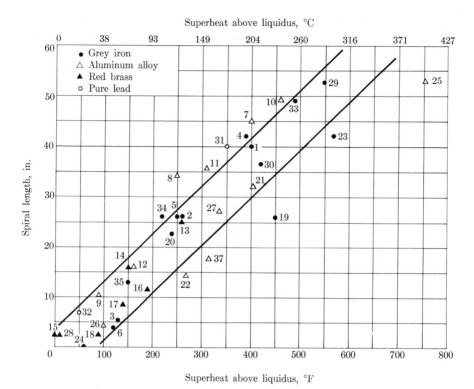

FIG. 5–5. Fluidity versus °F superheat above liquidus [4].

been noted in the alloys with pronounced film-forming tendencies, such as those caused by high aluminum contents; the fluidity of these alloys is lower than the general pattern (Fig. 5–5).

5–6 Suction-tube data. Early experiments by Portevin and Bastien [4] indicated that the fluidity of certain compositions, notably pure metals and eutectics, was greater at a given superheat than that of other compositions. More recently Ragone, Adams, and Taylor [5] developed a new method for measuring fluidity. A glass or metal tube is first dipped in the liquid metal at an angle. A stopcock connecting the upper end of the tube to a vacuum tank is then opened. The metal flows up the tube under the influence of the pressure head produced by the difference between atmospheric and tank pressures. The length of the metal column is taken as an index of fluidity just as in the fluidity spiral. The method has two advantages: it permits direct observation during metal flow, and it is somewhat simpler than pouring a sand-cast spiral. A possible disadvantage is that the thermal gradient and the nucleation conditions produced by the glass-mold wall are different from those arising in a sand mold. We shall shortly discuss this point as related to constitutional supercooling effects.

Data obtained by means of this apparatus confirmed the greater fluidity of pure metals and eutectics, particularly for lower melting-point materials such as aluminum, tin, zinc, and magnesium alloys. The data for the Al-Mg system [7], for example, show the following trend:

% Mg	Liquidus,		Fluidity at 50°C (90°F) superheat above liquidus, inches
	°C	°F	
0 (pure Al)	660	1220	19
5	630	1166	3
20	550	1022	4
33 (eutectic)	450	842	12

In analyzing these data the investigators developed a correlation with the liquidus-to-solidus temperature interval. The greater the liquidus-solidus interval, the lower the fluidity when measured at, or 50°C (90°F) above, the liquidus. Good evidence is offered that a *pure* metal stream stops flowing in a tube or spiral because it freezes off near the entrance, not at the end away from the crucible. Sectioning shows large columnar crystals near the point of metal entrance and finer crystals accompanied by centerline shrinkage voids away from the entrance.

5–7 Summary of spiral and suction-tube data. We may now attempt to reconcile the spiral data which indicate the importance of superheat above the liquidus and the suction-tube findings which place greater emphasis upon the liquidus-solidus interval of the particular alloy.

Pure metal. Recent experiments using the conventional spiral indicate that high-purity aluminum (99.99%) possesses much greater fluidity than "commercially pure" (99.0%) aluminum. There is definitely a difference in freezing mechanism between pure metals and eutectics on one hand and alloys with an appreciable freezing range on the other.

The following explanation is suggested. The pure metal freezes by columnar growth, which will begin near the gate end of the fluidity spiral, depending upon the superheat above the freezing point. In other words, solid metal will not start to build up at a given point along the mold wall unless the rate of heat transfer is sufficient to take care of both the superheat and the heat of solidification. Consider the fate of a solid layer attempting to grow near the ingate. If heat transfer to the layer from the liquid passing by is greater than the removal of heat by the mold, the layer will melt. Now, as the liquid passes down the tube its temperature will fall, and at some location, growth of a solid layer will begin. In a pure metal this layer will have a smooth interface, as discussed in Chapter 2, and thus no crystallites will fall into the stream. Eutectics exhibit a similar effect.

Alloys. In contrast, an alloy with a freezing range exhibits constitutional supercooling. It is suggested that the distribution coefficient k as defined in Chapter 2 is more important than the temperature range involved. In this case, as soon as a thin layer freezes the adjacent metal builds up rapidly in solute. As a result the freezing point of the layer is lowered, and freezing stops. Random crystallization in the stream will then prevail over columnar growth from the mold walls. As this crystallization progresses in the moving end of the stream, the metal becomes viscous. Finally the force required to move the stream is greater than the metal head, and the stream stops. In an alloy with a freezing range, the stream motion stops from the farther end rather than at the ingate.

Most cast alloys are of the second type, i.e., neither pure metals nor eutectics. In this group, therefore, the fluidity may be considered as related to superheat above the liquidus as a first approximation.

5–8 Application of fluidity data to casting problems. From a practical standpoint, it is immediately evident that to use our information about fluidity, we need to know the answers to the following questions. 1. How can the spiral data be applied to casting problems? 2. Since fluidity is a function of liquidus temperature, can the latter be predetermined?

Because the concept of "superheat above the freezing point" provides an approximate index of fluidity for many commercial alloys, it should be possible to use any one of a variety of alloys to pour the same casting satisfactorily, simply by maintaining constant superheat above the respective freezing points. This has been adequately proved commercially. It should be emphasized, however, that variation in pouring techniques or gating must be taken into account. For example, a drossy metal with an elaborate gating system must naturally be poured with a greater superheat.

To determine the proper pouring temperature for a new casting, i.e., one which has not been poured in any alloy, a pilot casting is necessary. The following empirical rules apply to the selection of test temperatures:

1. For fairly intricate castings of light sections ($\frac{1}{2}$ in. and under), test temperatures are usually from 300 to 500°F (167 to 278°C) above the liquidus.

2. For heavy castings such as machine bases, the range is 100 to 300°F (56 to 167°C) above the liquidus.

TABLE 5-2

DEPRESSION OF FREEZING POINT OF IRON PER WEIGHT
PERCENT FOR SEVERAL ALLOYING ELEMENTS

Element	Factor (additive)	Range tested
Carbon	120 (average)	Varies considerably (value given fairly good up to 1% C)
Manganese	9.0	0– 3.0%
Phosphorus	54	0– 0.7
Sulfur	45	0– 0.08
Silicon	14	0– 3
Nickel*	7.2	0– 9
Chromium	2.7	0–27
Nitrogen	162	0– 0.125
Aluminum	0	0– 1
Copper	9.0	0– 0.3
Molybdenum	3.6	0– 0.3
Vanadium	3.6	0– 0.3

* For more than 9% Ni, the factor is $(101 - x)/1.28$, where $x = $ w/o Ni.

5-9 Calculation of liquidus. If neither the thermocouple nor the optical-pyrometer technique is available for freezing-point determinations, an approximate calculation may be made for ferrous alloys from the data of Table 5-2. For example, the liquidus of a steel containing 0.30% carbon, 0.60% silicon, 0.50% manganese is estimated to be

$$2802 - (0.3)(120) - (0.6)(14) - (0.5)(9) = 2753°F (1512°C).$$

These data are based upon the fact that the lowering of the liquidus is brought about by the cumulative effect of the dissolved elements, and show good agreement with practice in a number of cases. It should be remembered, furthermore, that it is difficult to keep variations in pouring temperature to less than 100°F (56°C) when a series of castings is to be poured from the same ladle. An error of 20°F (11°C) is therefore not vital. For other alloys the available phase diagrams should be consulted.

REFERENCES

1. C. H. Desch, "Physical Factors in the Casting of Metals," *Foundry Trade Journal* **56,** 505 (1937).

2. C. M. Saeger and A. I. Krynitsky, "A Practical Method for Studying the Running Quality of a Metal Cast into Foundry Molds," *Trans. A.F.S.* **39,** 513–540 (1931).

3. W. S. Mott, R. H. Schaefer, and E. Cook, "Experimental Production of Pilot Static and Centrifugal Castings for the Armed Services: Part III—The Fluidity of Cast Alloyed Steels and Irons," NDRC Div. 18, OSRD No. 5634.

4. R. A. Flinn, W. A. Spindler, and W. B. Pierce, "A Revised Fluidity Spiral Test," *American Foundryman* (August, 1954).

5. A. Portevin and P. Bastien, "Fluidity of Ternary Alloys," *J. Inst. Metals* **54,** 45 (1934).

6. D. V. Ragone, C. M. Adams, and H. F. Taylor, "A New Method for Determining the Effect of Solidification Range on Fluidity," *Trans.* **64,** 653–657 (1956).

7. S. Floreen and D. V. Ragone, "The Fluidity of Some Aluminum Alloys," *Trans. A.F.S.* **55,** 391–393 (1957).

General References

Many articles on fluidity appear in *Transactions of the A.F.S., The Foundry,* and in British journals.

PROBLEMS

1. Criticize the statement, "A 3.5% carbon cast iron is more fluid than a 2.5% carbon malleable iron, and therefore lighter and more intricate sections can be cast." If you do not agree, explain why cast iron is chosen for delicate art work.

2. Explain the following data:

	High-purity Al 99.99% Al	Commercial pure Al 99.0% Al
Length of spiral poured at 1350°F (732°C)	27 in.	20 in.
Arrest due to heat of solidification	1220°F (660°C)	1218°F (658°C)

3. Suppose that your plant has been making a 0.3% carbon-steel part and that you have been asked to specify a new pouring temperature for the same part now to be made in 70 Cu 30 Zn brass. What is the simplest procedure?

CHAPTER 6

STRESS-STRAIN RELATIONS IN CASTINGS DURING COOLING AND HEAT TREATMENT

6–1 Introduction. The problem of obtaining a successful casting does not end with pouring metal at the proper temperature and rate into a mold with a well-designed gating and risering system. At this stage new problems may arise such as hot tears, injurious residual stresses,* as well as distortion during cooling, machining or heat treatment.

With this warning in mind, let us examine the situation from a positive point of view. We can apply our knowledge of the stress-strain relations existing in the casting at various temperatures, not only to avoid defects but to develop superior performance by producing residual stresses in directions opposite to the service stresses. In addition, a knowledge of plastic behavior at elevated temperatures enables us to properly control operations such as heat treatment or welding of complex assemblies.

It may seem that we are attempting a rather broad coverage by ranging in one chapter from topics such as hot tears, which occur on solidification, to the delicate distortions on the order of thousandths of an inch which may occur much later during machining. However, we shall see that these subjects and the other topics to be treated in this chapter are all basically related to stress-strain behavior at different temperatures. Therefore we shall first describe the plastic and elastic behavior of the casting and then apply the principles emerging from our discussion to the problems of hot tears and residual stresses in the order in which these defects arise in production. Our development will be patterned on the following outline: (1) stress-strain relations as a function of temperature; (2) hot tears (during solidification); (3) cracks and residual stresses which develop upon cooling below the solidus, heat treatment and stress relief.

6–2 Stress-strain relations as a function of temperature. Let us begin by considering the behavior of a metal bar as tension is applied, starting from the liquid state. We shall use as an example a bar of aluminum alloy 112F (7% Cu, 2% Zn, balance Al) (Fig. 6–1). For the time being, let us reserve our discussion of the *sources* of tension (which are abundant) until later. As we observe the change in behavior of the metal with falling temperature, we can distinguish five discrete stages.

---

* Also called internal or locked-up stresses.

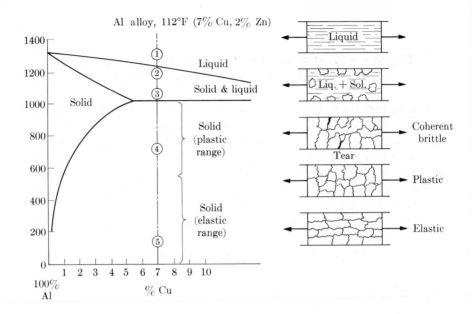

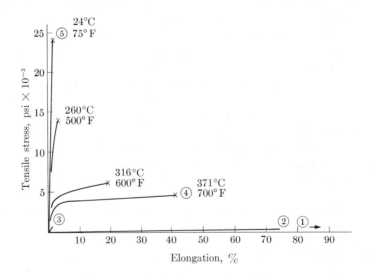

FIGURE 6–1

(a) *Completely liquid.* The metal follows whatever motion the end plates make. Obviously a tear cannot develop since it would be quickly filled by liquid. We also note that with falling temperature the liquid contracts, but this shrinkage is compensated for by liquid from the riser, as discussed previously (Chapter 2).

(b) *Mostly liquid with some solid.* From our earlier discussion (Chapter 2) we recall that solidification will take place at constant temperature in a pure metal or eutectic, while in other alloys solidification will occur over a temperature range. In either case, when the bar is stressed, adequate liquid is available to fill in any potential rupture.

(c) *Mostly solid with some liquid.* At a certain stage during solidification there emerges a connected network of solid crystals which has some strength, i.e., the metal becomes *coherent.* It has been variously estimated that this happens at the time when the metal is 50 to 90% solid, depending upon its crystallization pattern. Liquid films are still present, and the structure ruptures under very low stresses. The isolated patches of liquid metal are not able to fill in all the tears caused by tension. Thus we encounter our first defect, which is called the *hot tear* because it occurs at or just above the solidus. We shall observe specific x-ray evidence of this defect later. It should be noted in passing that if liquid films of low-melting impurities such as sulfides or phosphides are present, this type of rupture can take place below the solidus shown by the phase diagram for the *pure* alloy.

(d) *Solid: from below-solidus temperature to the "elastic" or cold-work range.* This is the *plastic* range in which ductility is high, and metal flow takes place at low stresses. Unless there exists a grain boundary network of a phase that is brittle at this temperature (such as a brittle compound), cracking does not occur. The metal grains tend to contract, but if they are restrained at the ends of the bar, they flow plastically. Furthermore, if the grains are deformed, they recrystallize. (This is the familiar "hot forging range" in which hot working operations are performed on wrought materials.)

Another feature of this range is the phenomenon of *creep* which is important to our later discussion of stress relief. If we apply a steady load to the bar, the grains will continue to flow or elongate as a function of time. We will see later that if a cold casting with residual stresses (represented by locked-up elastic strains) is heated to this range, the stresses will be relieved by plastic flow.

(e) *Solid: "elastic" range below recrystallization temperature.* There is really no sharp transition from plastic to elastic behavior, nor is there an exact recrystallization temperature. (Note the change in the stress-strain curves of Fig. 6-1.) One can see, however, that as the temperature is lowered, a point is reached at which the grains elongate and do not

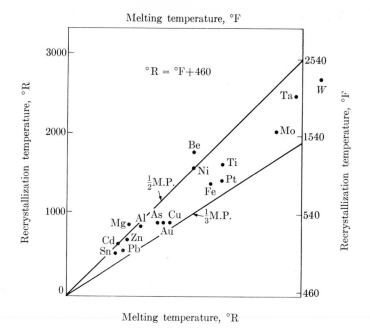

FIGURE 6–2

recrystallize and that the *flow stress,* i.e., the stress required for plastic deformation, increases. Now, the important characteristic of this range is that if we apply a stress below the yield strength, elastic strain develops instead of rapid relief of the stress by creep. The temperature of this transition from plastic to elastic behavior varies for different materials and is approximately one-half to one-third the melting point expressed in degrees absolute (Fig. 6–2). Thus we see that iron is plastic from 1000 to 2800°F (538 to 1538°C), while for aluminum the range is 500–1200°F (260 to 649°C).

The stress-strain curves for the different temperature ranges are summarized in schematic fashion in Fig. 6–1. Because the metal in stages (a) and (b) is predominantly liquid, curves (1) and (2) show no appreciable stress, and the strain is infinite. When the metal becomes coherent, it fractures with little deformation and at low stress as indicated by curve (3). Curve (4) shows marked plastic behavior at low stress, and curve (5) exhibits elastic deformation followed by limited plastic flow at high stress.

We can now proceed to discuss the specific data applying to the formation of hot tears, stage (c), and then deal with the problem of residual stresses which develop in range (e) due to the plastic deformation in range (d).

6–3 Hot tears: mechanism and experimental evidence. We shall first review the experimental evidence related to the formation of hot tears and then take up the variations in chemistry and mold design which can be used to prevent these defects.

Not until quite recently have investigators been able to provide a satisfactory explanation of the mechanism of hot tearing [1]. Although earlier investigators had determined stress-strain curves at different stages during cooling of cast tensile bars, their temperature measurements were later proved to be in error by several hundred degrees fahrenheit below the actual range [2]. Equally erroneous was their conclusion that castings were brittle far below the solidus. Later work [1] on aluminum alloys and on steels of different compositions established that hot tearing takes place between the temperature at which the metal becomes coherent and the end of freezing at the solidus.

Let us review the data for the aluminum-base alloy (4% Cu) first. A plate of dimensions $1 \times 6 \times 24$ in., approximately, is cast with chilled L-shaped end sections to provide restraint during cooling (Fig. 6–3a). Tensile strain therefore develops as the casting cools and contracts. Side risers are positioned at the center section to provide sound material for observation, except as noted later. Thermocouples are cast in place to measure the temperature of the center section, and the temperature difference between the surface and the central plane does not exceed 10°F (6°C). During cooling, the castings are x-rayed by means of 10-second exposures at intervals of 30 to 60 seconds. Upon examination of the films, the temperature at which tearing begins can be established quite accurately. These results are plotted in Fig. 6–3(b), along with the liquidus and solidus temperatures for this alloy. It is apparent that the hot tears are initiated before the section is solid. When the risers are reduced in size, tearing begins sooner because liquid feeding is unavailable at an earlier time during solidification.

A more extensive investigation of the same type was conducted for steels varying in carbon content from 0.03% C to 0.96%. In Fig. 6–4 the temperature for the beginning of hot tearing is plotted directly on the iron-carbon diagram. As in the aluminum alloy, hot tearing begins at or above the equilibrium solidus temperature* for all compositions within the limit of error of observation. This conclusion seems to be contradicted by several points on the graph, namely those marked 0.09 S, 0.12 S, 0.11 P, and 0.07 P. The sulfur and phosphorus content of these heats was deliberately increased over the typical 0.02% S and 0.02% P. Also the mold was changed by shortening the plate so that a melt of standard composition would not tear, since tearing would obscure the effect of the

---

* Note that segregation during actual freezing would lower the solidus temperature (Chapter 2).

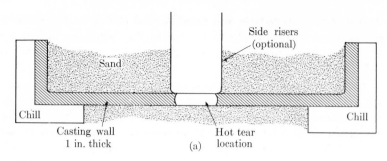

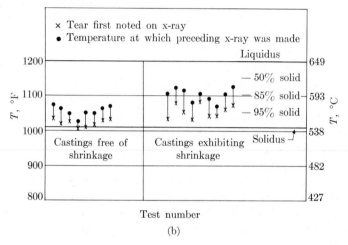

FIG. 6–3. (a) Cross section of hot-tear test casting (approx. ⅛ size). (b) Relation of temperature at which hot tearing begins in 96% Al-4% Cu castings to AlCu phase diagram and to casting soundness [1].

minor elements. The true solidus for these special melts was lowered by the presence of the low-melting sulfur- and phosphorus-rich phases. These heats correlate well with the generally accepted principle that high sulfur and phosphorus contents lead to more severe incidence of hot tearing.

Although we shall consider additional evidence in the next section, the critical experiments just described indicate that when hot tearing takes place, it occurs at or just above the solidus in a casting under tensile stress. It is important to emphasize that it is the combined effect of two factors which produces a tear: metal in the coherent state with liquid films present *and* a sufficient tensile stress *at this time*. Thus an alloy which is extremely sensitive to tearing may be cast without tears if the mold is designed so that no tensile stress occurs during the coherent

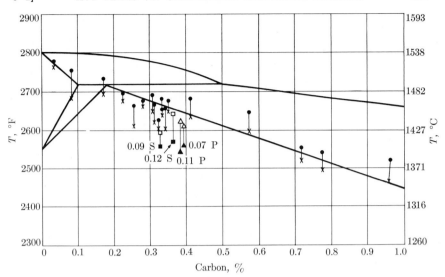

FIG. 6–4. Relation of temperature at which hot tears develop in steel castings to the iron carbon diagram. (Solid circles indicate that no tear is shown by radiograph; x indicates first radiograph showing tear.) [13].

(brittle) range. As an important corollary, a metal in which there is extensive final solidification at *constant temperature* will not hot tear in a casting of uniform cross section. There will be no tensile stress due to solid contraction at the time when the metal is in the period of final solidification, since the temperature is constant. We expect therefore to find eutectics and pure metals less susceptible to hot tears than alloys with a freezing range.

Let us now examine the evidence for this observation in various commercial alloy systems.

6–4 Relation of hot tears to chemical composition and phase diagram. (a) *Iron-base alloys.* The steel compositions discussed above freeze over a range of temperatures and, as expected, are susceptible to hot tearing in a casting design which imposes constraint. A coherent and brittle structure exists over a temperature range, and as the casting contracts during this period the solid is torn in regions where liquid films are present. The data of Fig. 6–5 provide additional evidence of interest. Tensile bars having a 3-inch diameter were cast and tested at different temperatures during cooling by applying stress to the bar in place in the mold. Note particularly that even an extremely low load is enough to fracture the bar until a skin at least $\frac{1}{4}$ inch thick develops. Thereafter the load-carrying capacity rises very rapidly as the remainder becomes

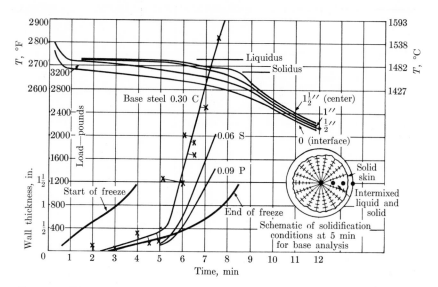

FIG. 6–5. The development of strength in various steels with increasing solidification [1].

solid. The effect of phosphorus and sulfur in delaying the time at which the bar develops appreciable strength confirms our earlier conclusion that these elements form films of low melting point.

Cast iron, ductile iron, and malleable iron may be considered next. In actual plant practice, hot tears are extremely rare in gray cast iron and ductile iron. In both of these materials which contain from 3 to 4% carbon and from 1 to 3% silicon, the final solidification, involving over half the structure, takes place at a practically constant eutectic temperature. Hence, in reasonably uniform castings, there is no tension due to contraction while the coherent, brittle structure is present. One may ask in passing why gray cast iron, which exhibits low elongation at room temperature, does not fail when stresses develop just below the solidus. The answer is that in the range from the solidus at 2060°F (1127°C) to the elastic range starting at about 1000°F (538°C), both white and gray cast irons possess appreciable ductility (of the order of 10%). Another factor tending to prevent hot tears in both ductile and gray iron is the *expansion* during final solidification which results from the reaction liquid → graphite + austenite. One interesting case of hot tearing in gray iron has been reported [3]. In this instance the phosphorus content was raised from 0.04% to 0.25%, and the ductility was measured in the range 1950°F to 1700°F (1066 to 927°C); it was found that the elongation was reduced by 50% and the tensile strength by 10 to 40%. These

TABLE 6–1

COHERENCE RANGE OF ALUMINUM ALLOYS
(After Sicha and Stonebrook [5] and Gamber [10])

Alloy number	Coherence range		Alloy number	Coherence range	
	°F	°C		°F	°C
43	15	8	B195	54	30
214	18	10	113	61	34
A132	27	15	319	66	37
C113	27	15	108	67	37
356	29	16	214	88	49
355	33	18	122	91	51
D132	34	19	212	92	51
333	36	20	112	99	56
A108	46	26	195	146	81
F214	49	27	220	176	98

effects can be ascribed to the presence of a low-melting ternary eutectic of iron phosphide, iron carbide and iron. Also, high phosphorus content should be avoided in the presence of molybdenum in alloy cast irons because a greater volume of a complex, low-melting phase is obtained which is rich in both phosphorus and molybdenum.

At the carbon and silicon contents of the malleable-iron composition range (2.2 to 2.5% C, 0.7 to 1.5% Si), the amount of eutectic is greatly reduced. For example, at 2.2% C and 1% Si, 20% of eutectic is obtained. In addition, these compositions are cast as white irons, and therefore the expansion due to graphitization does not occur during freezing; as a result these materials are susceptible to tears [4].

(b) *Aluminum- and magnesium-base alloys.* Interesting experiments [references 5 through 12] have been conducted in this field in addition to the work previously discussed. It has been considered [5] that the structure of aluminum alloys becomes coherent when 70% of the metal is solid. Then, from calculations based on phase diagrams and from actual measurement, the temperature range can be determined during which a given alloy is coherent (70 to 100% solid). These data, presented in Table 6–1 for most of the important sand-casting alloys, correlate well with practical field experience in all cases except the 220 alloy, which is less sensitive than the calculations would lead us to expect.

Similar work for other aluminum alloys as well as for magnesium alloys confirms these findings. A series of eight U-shaped test bars has been developed [10] with fillet radii ranging from a sharp corner to

TABLE 6–2

COHERENCE RANGE OF MAGNESIUM ALLOYS

Mg alloy	Al	Mn	Zn	Zr	Rare earths	Th	Hot-cracking rating
	(balance of composition is magnesium)						
EK 41A	—	—	—	0.6	4.0	—	3.0
EZ 33A	—	—	2.5	0.6	3.3	—	3.5
HK 31A	—	—	—	0.7	—	3.2	4.2
AM 100A	10	0.1	—	—	—	—	4.4
AZ 92A	9	0.1	2.0	—	—	—	4.9
HZ 32A	—	—	2.1	0.7	—	3.2	5.5
AZ 63A	6	0.15	3.0	—	—	—	7.5
ZK 61A	—	—	6.0	0.7	—	—	7.5

¾ inch. Alloys which do not crack when poured into the sharp-notched mold are given a rating of 1.0. The rating increases to 8.0 for the largest radius. The test results for aluminum alloys agree quite well with the data of Table 6–1.

The same technique applied to magnesium-base alloys gives the ratings listed in Table 6–2.

In general, the alloys containing rare earths exhibit less susceptibility to tearing because of the eutectic liquid which freezes at constant temperature during final solidification. By contrast, the alloys with poor ratings solidify over a long temperature range and have little or no constant-melting liquid during final solidification.

For completeness we should mention that the copper-base alloys exhibit, in general, only slight susceptibility to hot tearing. This characteristic is probably due to the fact that the tin-lead-zinc family of leaded bronzes contains a good deal of low-melting eutectic. Moreover, in the aluminum bronzes and the copper-zinc brasses the alloys freeze over a very narrow temperature range so that the temperature interval is very small while the coherent structure is present.

6–5 Relation of hot tears to mold constraints and to the interaction between sections. The two sources of stress causing hot tears are (a) differences in contraction of mold and casting and (b) differences in time at which contraction occurs at different locations within the casting. Let us consider each type separately.

(a) *Mold-casting interaction.* If a thin ring of metal is poured around a steel bar, a condition develops which is similar to the shrink fit of a hot ring on an arbor. Not only is hoop tension created by the contraction of the ring upon the bar, but in the early stages of cooling the bar is heated and expands.

This example has been selected because one of the most common causes of hot tears, called bore cracks in this instance, is the contraction around a core or a metal chill. It has been found that the compressive strength of a strong steel-molding sand can reach over 1000 psi at 2000°F (1093°C) [13], and thus exceeds the strength of the metal in the coherent range. It is not necessary that the casting surround a core completely. For example, a U-shaped section such as a long pipe with flanges creates the same condition. To alleviate this effect, both mechanical and chemical methods are used. Thus an insert may be rammed up in the mold and removed shortly after pouring, as soon as the casting skin has formed, to create a hollow space for mold expansion, or materials such as wood flour which burns out quickly may be added to the core mixture to provide for rapid collapse. Another method of relief is the use of green sand in place of core sand. Mold restraint can also operate through protruding gates or risers as indicated in Fig. 6–6(b and c). The twin gating system shown in Fig. 6–6(b) freezes rapidly and anchors the ends of the casting. The mold expansion places the casting in tension and tearing occurs at the hot spot. The risers in Fig. 6–6(c) act as anchors through which the mold expansion is transmitted. The central tear again takes place at the hot spot.

Another factor related to strength and heat capacity is the core density. A correlation of hot tearing with density [14] is graphically illustrated in Fig. 6–7, and an example of this type of tearing is shown in Fig. 6–6(a).

(b) *Differences in time of contraction at different parts of the casting.* Hot tears also result when contraction occurs at different periods during cooling and is not uniform throughout the casting. The differences, usually caused by interaction between light and heavy sections, can be aggravated by hot spots due to casting design, gating, risering, or shrinkage.

To present a simple example, let us consider the cracking of a thin flange attached to a massive rim (Fig. 6–8). In this case, the flange freezes completely shortly after pouring, and at the same time a strong shell of metal freezes at the sides and back of the rim. The temperature of the flange will continue to drop, whereas the temperature of the rim will stay practically constant because of the large mass of liquid metal behind the rim wall, which has yet to freeze. The flange therefore tries to contract, but is restrained by the rim shell. When the tensile strength of the flange is reached, tearing takes place. This type of interaction, which can occur on rectangular as well as circular shapes, is an important reason for eliminating light flanges and sharp edges.

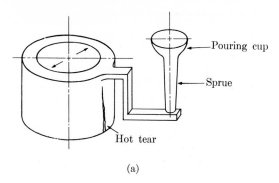

(a)

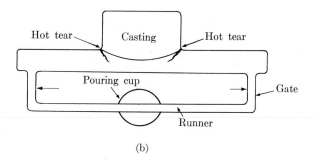

(b)

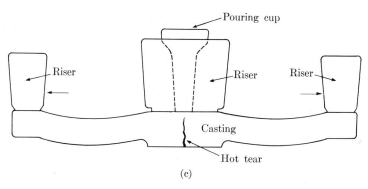

(c)

Fig. 6–6. Relation of hot tearing to mold-casting interactions. (a) Mold restraint due to casting cavity. (b) Mold restraint acting through the gating system. (c) Mold restraint acting through the riser system.

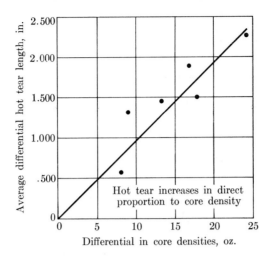

FIG. 6–7. Relationship between core density variation and resultant hot-tear variation [15]. (Core density was varied using different mixes.)

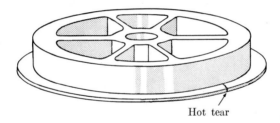

FIG. 6–8. Relation of hot tearing to conditions within the casting illustrated by the example of a thin flange attached to a massive rim.

6–6 Cracks, residual stresses, stress relief. We have now reached the point in our discussion at which we know how to produce a sound casting, free from hot tears, at a temperature in the plastic range just below the solidus. If we next develop an understanding of the interrelations of plastic and elastic strains and the temperature distribution during cooling, we shall possess a powerful means of controlling the final properties of the casting. At our option, we may then attempt to obtain either a pattern of helpful residual stresses or a stress-free casting. Furthermore, we may predict with confidence the type of stress pattern which will be produced during subsequent heat treatment or welding. The penalty for neglecting these basic relations clearly emerges from the long history of foundry technology, which is filled with reports describing how large cast-

TABLE 6-3

COEFFICIENTS OF THERMAL EXPANSION OF CERTAIN ALLOYS

Alloy	Composition	Coefficient of thermal expansion		Temperature	
		in/in · °F $\times 10^{-6}$	cm/cm · °C $\times 10^{-6}$	°F	°C
Steels	%C 0.30, %Mn 0.60, %Si 0.60	7.5	13.5	0–1300	−18–704
	%C 0.06, %Si 1.0, %Cr 18, %Ni 8	11.1	20	0–1800	−18–982
	%C 0.50, %Si 2.0, %Cr 18, %Ni 36	10	18	0–1800	−18–982
Cast iron	%C 3.15, %Si 2.16	7.5	13.5	0–800	−18–427
	%C 2.98, %Si 1.23, %Cr 1.61, %Ni 21	10.2	18.4	0–800	−18–427
	%C 2.3, %Si 1.0	6.6	11.9	0–200	−18–93
Copper alloys	%Cu 94.9, %Si 4.0	10	18	0–500	−18–260
	%Cu 85, %Zn 5, %Sn 5, %Pb 5	10	18	0–200	−18–93
Nickel alloys	%Ni 97	8	14.4	70–850	21–454
	%Ni 67, %Cu 29	8	14.4	70–850	21–454
	%Ni 64, %Cr 20, Balance Fe	9.4	15.1	20–1800	−7–982
Aluminum	%Al Bal., %Mg 1.0, %Si 12, %Ni 2.5 (A132), %Cu 1.0	10.3–12.0	18.5–21.6	−58–572	−50–300
	%Al Bal., %Mg 0.3, %Si 7 (356)	11.0–12.9	19.8–23.2	−58–572	−50–300
Zinc base	%Zn Bal., %Cu 2.5, %Al 3.5	15.4	27.7	68–212	20–100
	%Zn Bal., %Cu 0.1, %Al 4	15.2	27.4	68–212	20–100
Magnesium AZ92A-T4	%Mg Bal., %Al 9, %Zn 2	14.8		68–212	20–100

ings such as rolls, engine frames, car wheels, and gear blanks have failed explosively hours or days after casting while cooling under sand in a pit, resting on a shipping platform, or in actual service. Indeed, castings with undesirable residual stresses are far more dangerous than castings with easily visible hot tears; and even if dramatic failure does not take place, a casting such as a motor block that does not retain machined dimensions is certainly unsuccessful.

To discuss the problem thoroughly, we shall observe the following outline. (It is obvious that certain portions are covered in more detail in other texts, particularly those dealing with engineering mechanics, but for completeness some review is undertaken here.)

(1) Quantitative discussion of expansion and contraction during temperature change and phase transformation.

(2) The measurement of residual stresses.

(3) Case histories illustrating the development of internal stresses in actual castings during cooling caused by (a) thermal gradients, including the role of plastic deformation, (b) phase transformation.

(4) Control of residual stresses.

(5) Stress relief: (a) rate, (b) commercial cycles for various materials.

6–7 Quantitative discussion of expansion and contraction. To understand the origin and control of stresses produced during cooling, it is necessary to keep in mind only two simple rules.

(a) Metals expand on heating and contract on cooling, but this behavior is altered when a phase transformation takes place.

(b) Metals are, in general, weak and plastic above the recrystallization temperature, but stronger and elastic at lower temperatures. (However, you will recall that a range of combined brittleness and low strength is encountered just above the temperature of *complete* solidification.)

Since these rules are obviously of little use without quantitative substantiation, we shall now provide the relevant data.

(a) *Contraction and expansion.* In the solid state, the coefficient of expansion (and contraction) of practically all steels, malleable irons, and cast irons lies between 6×10^{-6} in/in·°F and 12×10^{-6} in/in·°F (Table 6–3). When the structure of the iron is body-centered cubic (ferritic), the coefficient is about 7.6×10^{-6} in/in·°F even if substantial amounts of carbide, graphite, or both are present, e.g. 3% carbon. When the structure of the iron is face-centered cubic (austenitic), the coefficient is approximately 11×10^{-6} in/in·°F. The transformation of austenite to ferrite and carbide in the elastic range is accompanied by considerable expansion, and depends both upon composition and the temperature of

TABLE 6–4

TRANSFORMATION EXPANSION IN STEEL*

(Transformation temperature was controlled by quenching from
1600°F (871°C) into bath at indicated temperature.)

Composition				Expansion (10^{-4} in/in) at transformation temperature, °F(°C)				
%C	%Si	%Mn	%Mo	400(204)	700(371)	800(427)	900(482)	1200(649)
0.30	0.31	1.49	0.61			56		39
0.44	0.21	0.72	0.0	100	45		42	33
1.04	0.17	0.29			33		17	11

* See reference 15.

transformation, as shown in Table 6–4. Note particularly that the transformation of austenite to martensite at 700 to 400°F (371 to 204°C) is accompanied by an expansion of 100×10^{-4} in/in, which is equivalent to the expansion due to a temperature change of 1000°F (556°C) [15]. This is particularly important in alloy castings, as will be discussed later. The coefficients of other important alloys are listed in Table 6–3. In passing, we wish to note that the coefficients of some aluminum alloys and austenitic cast irons exhibit a considerable degree of similarity and that therefore a press fit of these materials may be expected to stay tight over the elastic range even with heating to, say 400°F (204°C). Engineers use this characteristic to advantage in placing cylinder liners and valve guides of austenitic cast iron in aluminum heads and blocks, to minimize differential expansion.

(b) *Plastic behavior.* We have already discussed the ductile behavior of most metals at elevated temperatures. Another important feature is the tendency of metals in the plastic range to continue to deform or creep at constant load. In the case of steel, for example, if a bar is loaded at 1600°F (871°C) to 25% of its tensile strength, it will continue to elongate with time, while a similar bar loaded at 70°F (21°C) to 25% of its tensile strength (at 70°F) will not exhibit appreciable additional deformation with the passage of time.

6–8 Measurement of residual stresses. Before analyzing the development of residual stresses in an actual casting, we wish to review briefly various methods of measuring internal stresses, since such a review may help the reader to understand the nature of these phenomena. When a cylindrical steel bar of cross section 1 in² is gripped at its ends in a tensile testing machine, it will, of course, elongate under tensile loading. If two reference marks are placed 1 in. apart on the surface of the bar before stressing and stress is then applied, the distance separating the marks will change. Thus, for example, under a load of 30,000 lb, the distance between the marks will increase to 1.001 in., since

$$\text{strain} = \frac{\text{stress}}{30,000,000} \text{ in/in,}$$

within the elastic range of steel. When the load is removed, the distance between the marks will return to 1.000 in. Similarly, if a saw cut is made to isolate the gage section from the remainder of the bar instead of unloading the gage length by reversing the controls, the test length will revert to 1.000 in.

Now imagine that the test bar under the same stress of 30,000 psi is not clamped in the machine, but is a light section of a huge casting composed of the testing heads and base of the machine. Let us suppose further that the stress* in the bar is unknown to the observer. To determine the stress, he could place reference marks at desired locations under stress and then unload the section by cutting it away from the remainder. In our case, reference marks placed 1.000 in. apart under the "unknown" stress of 30,000 psi tension would be 0.999 in. apart after unloading. This mechanical method is one way of determining residual stresses. Small holes are accurately drilled and the distance between them is measured, and then the section is cut away for remeasurement.

Recently "SR4" gages have found wide use. This method depends upon the change in electrical resistance with the change in the length of a wire. A fine wire is cemented to the casting, and the wire resistance is measured; then the section is cut free and the wire resistance is remeasured. The change in electrical resistance is related to the change in strain by a calibrated instrument. When the direction of principal stress is not known, one must resort to the Mohr circle analysis (discussed in texts on mechanics) which requires that gages be placed in three directions. Note that in both these methods, the stress is experimentally determined without regard to section shape.

6–9 Case histories. To illustrate the problems arising from the presence of internal stresses, let us consider first the elementary case of a chilled iron railroad car wheel, which involves only stresses due to thermal gradients and demonstrates the effect that plastic deformation at high temperatures has on the development of residual stresses in the elastic range (Case I). Then we shall discuss the more complex case of a centrifugally cast bimetal roll involving stresses caused by phase transformation (Case II).

---

* In this illustration we assume uniaxial stress in the direction of the gage marks.

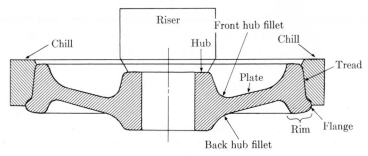

FIG. 6–9. Cross section of a 750-lb car wheel.

TABLE 6–5

RESIDUAL STRESSES IN MOLD-COOLED,
PITTED, AND AIR-COOLED WHEELS

Type of cooling	Radial stress on front hub fillet, psi	Circumferential stress on tread, psi
Mold- or air-cooled	32,200 (tension)	10,250 (compression)
Pitted after hot shake-out	645 (compression)	1,215 (compression)
Equalized 1450°F (788°C); accelerated hub cool	27,000 (compression)	

Case I. A typical mold for a chilled iron car wheel is shown in Fig. 6–9. In commercial practice, wheels are given a very careful three-day stress relief, i.e. they are removed from the molds at an average temperature of 1600°F (871°C) and placed in insulated pits. An obvious question suggests itself: Why does one not replace this slow cooling which results in a tensile strength of 33,000 psi in the plate section, by air cooling or quenching to obtain a fine pearlite or bainite formation which in turn would provide a plate strength of 60,000 psi. The answer is simple. It has been established experimentally that a wheel air-cooled from the mold either exhibits a high residual-stress pattern similar to that shown in Table 6–5 or shatters if the stresses are still higher.

Referring again to Fig. 6–9, we can see that the reason for the choice of this particular mold design is that it provides an ideal pattern of directional solidification toward the riser. The chiller, due to the area factor of 5:1 previously discussed, more than compensates for the heavier rim section. A temperature survey shortly after solidification (50 minutes after pouring) will show the hub region at 1800°F (932°C), the plate at 1500 to 1400°F (816 to 760°C) and the tread at 1400 to 1350°F (760 to 732°C). The plate and rim must have contracted more than the hub, but *no appreciable residual stress results at these tempera-tures because the iron is hot and plastic.* The result of the differential contraction of these two zones is plastic rather than elastic strain, and hence no appreciable residual stress is present in this temperature range. In other words, the creep rate of this material at 1400°F is rapid for light loads.

This temperature gradient persists if the wheel is air-cooled into the elastic range. For example, when the plate is at 800°F (427°C), the hot hub is at 1100°F (593°C); we then have the reverse of the case of a hot ring on a mandrel. Here the mandrel (the hub) is hotter than the ring by 300°F (167°C) and contracts away from the ring as the pair cools toward room temperature. The rim of the wheel contains residual cir-cumferential compression, but more importantly, the plate is in radial tension. Under rapid cooling the plate fails with a circumferential crack at the hub fillet, and the crack may then progress to the tread. Even if cracking does not occur during cooling, the casting obtained by this method would be unsatisfactory. Under service conditions, the heating of the tread by the brakeshoe can produce a severe thermal differential between rim and hub, leading to added radial tension. This additional stress might easily produce rupture in service.

By contrast, millions of practically stress-free wheels have been placed in service, with an excellent safety record. These wheels are removed from the molds after the same length of time that is proposed for the air-cooling cycle, with the tread at 1400°F (816°C) and the hub at 1800°F (932°C). Groups of 21 wheels are placed in a well-insulated, low-heat capacity pit (unfired), and the slow cooling rate of this assembly (10°F/hr) permits equalization of temperature *in the plastic range* (above 1000°F or 538°C). Since creep occurs rapidly in this zone, any elastic strain is relaxed to plastic strain, and the residual stresses are kept below 1500 psi, Table 6–5.

Case II. Now let us consider the case of the experimental bimetal roll illustrated in Fig. 6–10 [16]. In this case, the outer layer is composed of martensitic white cast iron and the interior of pearlitic gray cast iron. The casting is made centrifugally, with the outer surface being cast against a chiller. The alloy iron, 4.5% Ni, 2% Cr, is poured first, and after it has solidified the gray iron backing is poured. The advantages of bimetallic construction are greater toughness and excellent machin-ability of the bore.

In rolls of this type, cracks were often encountered during slow cooling several days after casting. These cracks were observed to occur in a radial direction through the chilled layer only, in the temperature range of about 400 to 700°F (204 to 371°C). It was evident that the mecha-nism involved here could not be the same as that operative in Case I

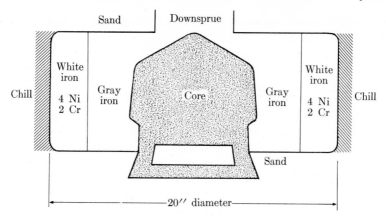

FIG. 6–10. Cast bimetal grinding roll [12].

because the roll was stripped from the mold shortly after pouring and the temperature was equalized in the plastic range, above 1400°F (760°C). Two paradoxical pieces of evidence soon developed. First, although the cracks were visible at the time of their occurrence and were obviously caused by hoop tension, they had closed by the time the casting reached room temperature and could then be detected only by magnaflux examination. Secondly, stress analysis of good rolls showed the existence of hoop *compression* rather than of hoop tension at the surface of the roll.

Since thermal gradients were eliminated as an important factor, the only other possible source of differential expansion was the difference in structure between the outer rim and the center. In both regions, the matrix is austenitic at high temperatures; however, the unalloyed center transforms to pearlite at temperatures in the range of 1400 to 1300°F (760 to 704°C), whereas the alloyed rim does not start to transform until the temperature has dropped to approximately 400°F (204°C) at which point martensite formation begins.

Transformation of the core to pearlite involves an expansion which tends to cause hoop tension. However, such stresses relax by creep at the high transformation temperature. Once the core has transformed to pearlite, it contracts at a rate of 7.6×10^{-6} in/in·°F until room temperature is reached. The rim, on the other hand, is still austenitic and contracts at a rate of 11×10^{-6} in/in·°F until the start of martensite formation (400°F). Between 1000 (538°C) and 400°F a differential strain of 2160 microin./in. can develop, and, using an average modulus of 20×10^6 psi, we clearly see that such a strain would result in a stress of over 40,000 psi, which is well beyond the tensile strength of the rim. Once the martensitic point is reached, expansion due to the martensitic reaction

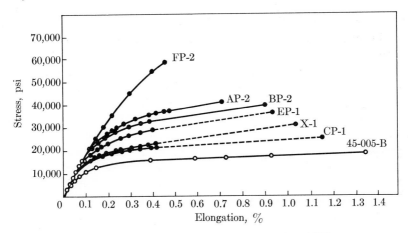

FIG. 6–11. Stress-strain curves for cast iron [17].

begins, cancels out the residual tension, and results finally in hoop compression. If cracking is to occur at all, it takes place before a temperature of 400°F is reached.

The cracking can be completely eliminated by using a 14% Ni, 6% Cu austenitic gray cast-iron center. In this case, the center has the same coefficient of expansion as the rim before the onset of martensite transformation, and rim tension does not develop.

6–10 Control of residual stresses. When adverse residual stresses are encountered in a casting, they may be reduced or changed in direction by control of the cooling period after pouring, or they may be altered by later heat treatment. We shall first consider an example of reversal of direction and then examine the data on rate of stress relief.

It was desired to perform an actual service test of experimental car wheels of the design previously discussed but with an unchilled tread of a martensitic gray iron structure obtained by quenching. In preliminary experiments the wheel was cast without a chiller, removed from the mold above 1400°F (760°C), and the tread was spray-quenched. This technique produced the desired structure, but the stress pattern resembled that of the air-cooled chilled wheel in that the contraction of the hub from the rim after quenching caused high radial tension in the plate that made the wheels obviously unsafe for service.

To remedy this condition, the dry-sand hub core was punched out as soon as the wheel had solidified, and a compressed air jet was injected to cool the bore preferentially. When the wheel was then removed for quenching, the hub was at 1200°F (649°C), while the tread and plate were above 1400°F, and hence a timed immersion quench of the entire wheel hardened only the plate and tread sections. Residual stresses in radial compression were obtained because the hub had been cooled before the tread.

An interesting corollary is represented by the stress-strain curves of Fig. 6–11 [17]. Although the high tensile strength (60,000 psi) of the quenched-plate structure may seem desirable, it is, in fact, unsatisfactory in a stress-free wheel because it will not permit the minimum requirement of 0.7% elongation in the plate that experience has shown to be necessary for resisting thermal stresses caused by heating of the rim. On the other hand, ferritic gray iron at only 30,000 psi is capable of the required elongation in a stress-free structure and hence is satisfactory. The quenched wheels prepared by the procedure just outlined had satisfactory effective elongation only because of the high residual stress in radial compression. (Before a tensile stress can develop in the hub fillet, the residual compression has to be overcome.)

6–11 Stress relief. We have discussed methods designed to prevent residual stresses by removing thermal gradients while the casting is in the plastic range, as well as procedures for controlling residual stresses by developing proper thermal gradients. In steel, malleable iron castings, and in many other alloys, the castings will receive subsequent heat treatment which can eliminate the stresses present in the as-cast condition. However, if severe gradients develop during cooling from the heat-treatment temperature, another set of residual stresses may result. Stresses may also be produced by welding and weld repair.

Hence stress relief treatment may be required to remove stresses of three different origins: (1) as-cast stresses, (2) residual stresses resulting from a heat treatment such as quenching, and (3) stresses produced by welding. All three types may be discussed from the same point of view because in each case the basic requirement is to convert elastic strain to plastic strain. To illustrate this point and to describe one method of obtaining stress relief data, we shall discuss the rate of stress relief, using the relaxation machine developed by Nadai [18] (Fig. 6–12), and then take up actual stress relief cycles.

The machine consists of three essential parts as shown in Fig. 6–12: the furnace, the loading mechanism, and the extensometer control system. A test specimen is held by threaded grips in an electric furnace, B, and heated to any desired temperature. The temperature controller, D, maintains the specimen at constant temperature for the duration of the test. The specimen is stretched a predetermined amount by applying a tensile stress through the lever arm, A, by means of G and H. The elongation is measured with the extensometer, E, and the load with the dial, J. The machine is then set for automatic operation. As the specimen tries to stretch plastically (creep) because of the load and temperature condi-

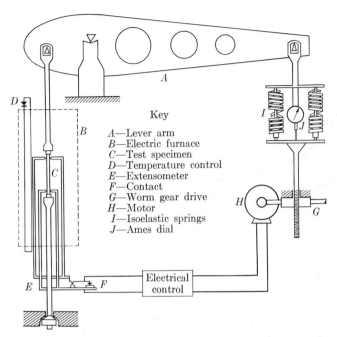

FIG. 6–12. Relaxation machine [18].

Key

A—Lever arm
B—Electric furnace
C—Test specimen
D—Temperature control
E—Extensometer
F—Contact
G—Worm gear drive
H—Motor
I—Isoelastic springs
J—Ames dial

tions, the extensometer contact, F, closes and starts the motor, H, which reduces the applied stress.

Due to sensitivity of the contacts, the loaded specimen does not actually elongate in practice. For example, let us assume that the specimen tries to stretch 10 microinches plastically during a one-minute interval. As soon as the slightest elongation occurs, the pressure on the contact causes a corresponding reduction in load and consequent removal of elastic strain. In the interval just discussed, the specimen remains at constant length, the load is reduced, and elastic strain is converted to plastic strain.

Typical data showing the decrease in elastic strain, and hence in the residual stress, as a function of time at a given temperature are presented in Fig. 6–13 for 0.25% C cast steel at 1020°F (549°C). To obtain these data, two values of initial stress were selected, approximately 12,000 and 6600 psi, respectively. Higher values were not chosen because of the decrease in yield strength with temperature (Fig. 6–14). In other words, the alloy is already in the plastic range, in which elastic strain is rapidly converted to plastic strain. Note, for example, that the stress falls from 12,000 psi to 10,000 psi in just a few minutes. Thus, in this temperature range, the determination of the yield strength itself is related to the speed of loading. Also, for alloys in which a change in the structure of the material such as tempering or overaging may occur on heating, the

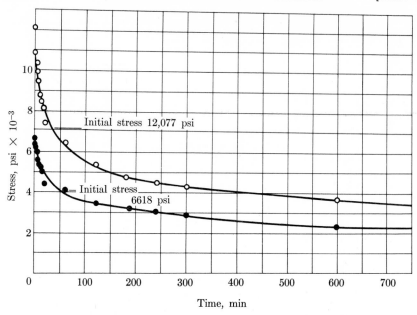

Fig. 6–13. Representative stress-time curves obtained by the constant gage-length method for 0.25% C cast steel tested at 1020°F (549°C). (After Rominski and Taylor [18].)

yield strength will depend upon the time at a given temperature before testing. This explains why in Fig. 6–14 the yield strength of 356T6, an aluminum alloy with a very fine precipitate, falls so rapidly with temperature.

We may correlate the temperature of stress relief and time at temperature [19], using data for cast steel, as shown in Fig. 6–15. We find empirically that these and other data yield a straight-line graph when the rate of stress relief at a given time is plotted against the stress remaining at that time (Fig. 6–16). Hence the rate of stress relief can be expressed by an equation of the type

$$\text{Rate of stress relief} = \frac{dS}{dt} = KS^n, \qquad (6\text{--}1)$$

where

S = stress at time t,

t = time under load at stress relief temperature,

K, n = constants for a given material at temperature T.

Taking logarithms of both sides, we obtain an equation of the form $y = mx + b$; i.e., a straight-line relationship:

$$\log\left(\frac{dS}{dt}\right) = n \log S + K. \qquad (6\text{--}2)$$

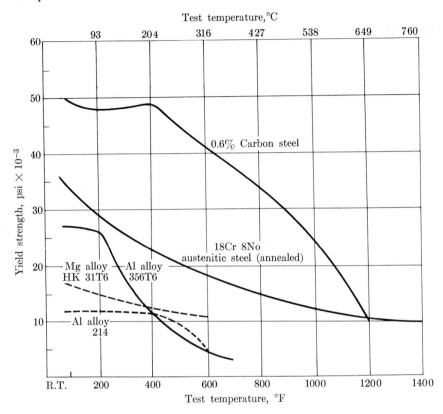

Fig. 6–14. Effect of temperature upon yield strength.

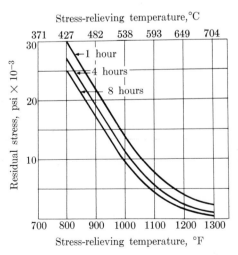

Fig. 6–15. Effect of stress-relieving temperature and time on the residual stress exhibited by a steel of yield strength 50,000 psi (original stress = yield strength).

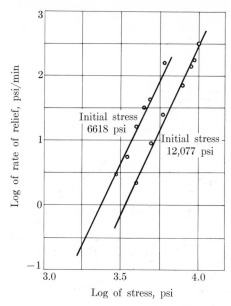

Fig. 6–16. The rate of relief as a function of stress level and initial stress for cast 1025 steel tested at 1020°F (549°C). (After Rominski and Taylor [18].)

If we integrate Eq. (6–1), using S_i to denote the original stress at $t = 0$ and S to denote stress at any time t, we obtain

$$S = [(1 - n)Kt + S_i^{1-n}]^{1/(1-n)}, \qquad (6\text{–}3)$$

and we see that the remaining stress, S, is not a linear function of time, but decreases in complex fashion. The data of Fig. 6–15 indicate that for steel in the temperature range from 800 to 1000°F (427 to 538°C) an increase of 60°F (33°C) in temperature is equivalent to an eightfold increase in time at the stress-relieving temperature.

6–12 Selection of temperature and time cycles for stress relief. From the preceding discussion it should be obvious that if stress relief is our only concern, we can develop the shortest cycle by heating to the maximum temperature which may be employed without incipient melting, warping, or excessive scaling. The casting is held at temperature, allowing plastic flow to relieve elastic strain, and then cooled slowly enough to prevent the development of appreciable thermal gradients. This type of full anneal and stress relief is widely used whenever it is desired to soften castings for better machinability as well as to relieve stresses.

On the other hand, when a casting has been heat-treated for the purpose of developing high hardness and strength, it is evident that the use of

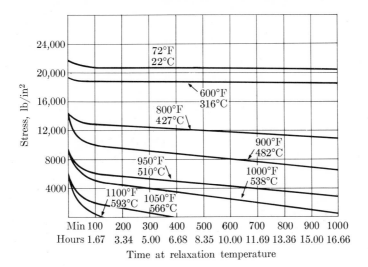

FIG. 6–17. Stress relief of cast iron (2.72% C, 1.97% Si, 0.51% Mn, 0.08% S, 0.14% P) at various temperatures [20].

a subsequent high-temperature stress-relief cycle would reduce the strength. It is not our aim here to review the metallurgy of different treatments such as age-hardening and martensitic transformation which are discussed in many metallurgical texts. We may simply say that all these treatments develop high strength by the dispersion of particles of a critical size from one phase to another phase. When this structure is heated to the stress-relieving temperature, the particles coalesce and their strengthening effect is drastically reduced. For the aluminum alloy 356T6, the yield strength at 70°F (21°C) would be lowered by 25% after a stress-relief cycle at 700°F (371°C). To avoid this impairment of a desirable property, it is better to conduct the quenching treatment in a way that will minimize or prevent the development of residual stresses. Thus, for example, using a spray mist or hot water instead of cold water for the quenchant minimizes temperature gradients and residual stresses. When castings are to be used at elevated temperatures, a stabilization treatment is employed which involves heating to about 50°F (28°C) above the service temperature, to minimize distortions and structural changes at the service temperature.

Data for the stress relief of gray irons of different compositions developed by means of the relaxation machine [18] are presented in Figs. 6–17 and 6–18. The specimens were loaded to reach a total strain of 0.3% at the start of the test, resulting in the various starting stresses shown at zero time. The data indicate that very little stress relief takes place below 900°F (482°C) and thus show that the old practice of

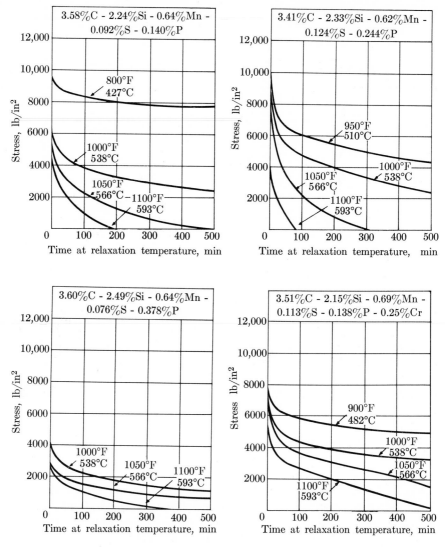

FIG. 6–18. Stress relief of various cast irons [17].

achieving stress relief by allowing castings to age at room temperature has little merit. Also, the role of chromium in decreasing the rate of relaxation is illustrated by the iron containing 0.25% Cr. Some recommended stress-relieving cycles are summarized in Table 6–6.

In general, copper-base castings do not require stress-relieving treatments since their relatively high conductivity leads to low thermal gradients in the plastic range. Aluminum sand castings also have rela-

TABLE 6–6

RECOMMENDED STRESS-RELIEVING CYCLES

Material	Temperature		Time, hour per inch of thickness
	°F	°C	
Carbon steels	1100–1250	593–677	1
Carbon molybdenum steel (0.5% Mo)			
Less than 0.2% C	1100–1250	593–677	1
0.2 to 0.35% C	1250–1400	677–760	3–2
Chromium molybdenum steel			
2% Cr, 0.5% Mo	1350	732	2
5% Cr, 0.5% Mo	1375	746	3
9% Cr, 1% Mo	1400	760	3
Chromium-nickel stainless steel (18–8) type	1500	816	2
Gray iron, ductile iron, malleable iron	900–1100	482–593	5–1

tively low residual stress levels for the same reason. Most aluminum- and magnesium-base alloys that are used in severely stressed applications are heat treated for age hardening, and therefore the relief of as-cast stresses is not a problem.

Before leaving the subject of stress relief, we wish to emphasize that heating to the stress relieving temperature must proceed at a rate which does not develop thermal gradients in the casting. It is possible to crack castings by uneven heating if the stresses due to the temperature gradient exceed the tensile strength. Similarly, warping may take place if the stresses exceed the yield point. Note that the actual stress required here may be much lower than the test-bar strength because of residual stresses and stress concentrations due to design. The allowable heating rate is also a function of the arrangement of castings in the furnace. For example, in an actual case of stress relieving car wheels of the design shown in Fig. 6–9, three wheels were stacked closely together in a furnace. The radiation from the elements in the walls, top and bottom, heated only the rim of the center wheel, leading to radial tension in the plate section and eventual rupture. The other two wheels were heated more uniformly since the hub and plate sections were exposed, and no failure was encountered.

Another question which often arises in stress relief concerns the rate of cooling. Again, there is no fixed rate; the only criterion is that the thermal gradient in the casting should not be large enough to cause plastic flow. It is usually possible to accelerate the cooling and tolerate greater thermal gradients once the elastic range is reached because of the concomitant increase in yield strength.

6–13 Relation between hot tears and residual stresses. Some confusing errors have developed in the literature concerning the relation between hot tears and residual stresses. The mold restraints which may be present in the stage of solidification in which the metal structure becomes coherent can cause hot tears, but will not lead to residual stresses because the metal is plastic. Also, there is an interesting inverse relationship between stresses causing hot tears and residual stresses arising from the interaction between casting sections. Note, for example, that the flange of the wheel shown in Fig. 6–8 will hot tear due to tension at elevated temperatures. However, if the flange does not tear, it will have been strained plastically in hoop tension in the plastic range because it cools faster than the rim. There is no appreciable elastic strain, but a thermal gradient exists as the rim and flange enter the elastic region. During cooling in the elastic range, the hotter rim will undergo a greater contraction than the flange, and hence the flange will be subjected to hoop compression as both reach ambient temperature. The residual stress is therefore opposite in sign to the stress causing hot deformation or tearing.

6–14 Weldments. We may briefly discuss the stresses arising from welding, to indicate that the same principles apply as for casting. Let us take the simple case in which a circular button of weld metal is applied to repair a void or to build up a surface of another composition such as a hard-facing alloy. During welding, there is, first of all, the simple local heating effect of the casting itself. The metal surrounding the weld attempts to expand and is restrained by the surrounding colder metal. If the thermal gradient is sufficiently high, the hot metal is deformed in compression. Next, a pool of liquid metal forms in the weld itself. No stress in the pool can develop while a large portion is liquid, but when the metal reaches the coherent range, hot tearing can occur. Hence, to minimize hot tearing, many welding-rod materials are made of eutectic compositions. After solidification, we have a hot, plastic, button-shaped region of the weld metal and, nearby, base metal surrounded by cold, elastic material. The residual-stress level at this point is low because of the plasticity of the weld. However, upon cooling through the elastic range to room temperature, the button will contract far more than the surrounding cold material. This leads to high biaxial tension in the weld button, similar to the tension found in the stretched skin of a drum. To alleviate this condition a number of remedies may be considered.

The simplest one is to use a very ductile metal in the weld which will flow plastically rather than crack under the tensile stress developing during cooling. The remaining elastic strain in the weld can be removed by a stress-relieving treatment or, more crudely, by peening the weld to cause compression. A better method is to preheat the casting to the plastic range before welding. Admittedly, a temperature gradient will develop between the weld and the base metal, but if this gradient is removed by conduction before the casting reaches the elastic range, no residual stress will result. Thus, when the casting cools slowly after welding, no additional heat treatment (postheat) is needed. With complex castings in which critical sections cool differentially, postheating and controlled furnace cooling are desirable.

Another problem in welding can develop if the composition of the rod being used is markedly different from that of the casting. This case requires the application of the principles discussed in the transformation expansion of the bimetal roll.

In general, most welding problems, including that of the weld assembly of several castings, can be analyzed quite simply by considering the weld itself as a casting, subject to the constraints of the nearby sections.

6–15 Summary. It may be helpful to review the entire thermal history of the casting.

Stage 1. Liquid metal is poured with sufficient superheat into a well-designed gating system, to produce a satisfactory surface.

Stage 2. The casting begins to solidify, and the risers provide liquid metal to compensate for the liquid-to-solid contraction of the casting.

Stage 3. The casting becomes coherent in the outer layers. Tearing is prevented either by selecting a composition with a long constant-temperature period of solidification or by removing mold restraints.

Stage 4. The casting is in the plastic range, and if thermal gradients are eliminated before the elastic range is reached, no residual stresses will develop in Stage 5.

Stage 5. The casting cools through the elastic range.

Stage 6 (optional). Stress relief and heat treatment.

REFERENCES

1. C. G. ACKERLIND, H. F. BISHOP, and W. F. PELLINI, "Metallurgy and Mechanics of Hot Tearing," *Trans. A.F.S.* **60,** 818 (1952).

2. H. HALL, "The Strength and Ductility of Cast Steel During Cooling from the Cast State in Sand Molds," Part I, American Iron and Steel Institute Special Report No. 15, pp. 65–93 (1936); also Part II, Special Report No. 23, pp. 73–86 (1938).

3. J. C. HAMAKER, JR., and W. WOOD, "Influence of Phosphorus on Hot Tear Resistance of Plain and Alloy Gray Iron," *Trans. A.F.S.* **50**, 502 (1952).

4. E. A. LANGE and R. W. HEINE, "A Test for Hot Tearing Tendency," *Trans. A.F.S.* **50**, 182 (1952).

5. W. E. SICHA and E. E. STONEBROOK, "Correlation of Cooling Curve Data with Casting Characteristics of Aluminum Alloys," *Trans. A.F.S.* **57**, 489 (1949).

6. D. C. G. LEES, "The Hot Tearing Tendencies of Aluminum Casting Alloys," *J. Inst. Metals,* **72**, 343 (1946).

7. A. R. E. SINGER and P. H. JENNINGS, "Hot Shortness of Some Aluminum-Iron-Silicon Alloys of High Purity," *J. Inst. Metals* **73**, 273 (1947).

8. A. R. E. SINGER and J. A. COTTRELL, "Properties of Aluminum-Silicon Alloys in the Region of the Solidus," *J. Inst. Metals* **73**, 33 (1947).

9. S. I. SPECTOROVA and T. V. LEBEDEVA, "Determination of Hot Shortness of Aluminum Alloys and Magnesium Alloys," Department of the Secretary of State of Canada, Foreign Languages Div., Bureau of Translation, No. 13561 (Russian) (BVP) May 4, 1955.

10. E. J. GAMBER, "Hot Cracking Test for Light Metal Alloys Castings," *Trans. A.F.S.* **67**, 237 (1959).

11. R. A. ROSENBERG, M. C. FLEMING, and H. F. TAYLOR, "Non-Ferrous Binary Alloys' Hot Tearing," *Trans. A.F.S.* **68**, 518 (1960).

12. R. A. DODD, W. A. POLLARD, and J. W. MEIER, "Hot Tearing of Magnesium Casting Alloys," *Trans. A.F.S.* **65**, 100–117 (1957).

13. H. L. YORK, "Report of Progress of Sand Research on Steel Sand Mixtures at Elevated Temperatures," *Trans. A.F.S.* **47**, 805–830 (1939).

14. C. H. WYMAN, "Hot Tears in Steel Castings," *Trans. A.F.S.* **50**, 152 (1952).

15. R. A. FLINN, J. A. FELLOWS, and E. COOK, "A Quantitative Study of Austenite Transformation," *Trans. A.S.M.* (1944).

16. C. R. WILKS and E. COOK, NRC Report 63A (1945).

17. R. A. FLINN and R. J. ELY, "Stress Determinations in Cast Irons for Railroad Service," *Trans. A.S.T.M.* (1950).

18. E. A. ROMINSKI and H. F. TAYLOR, "Stress Relief and the Steel Casting," *Trans. A.F.S.* **51**, 709–731 (1943).

19. C. R. JELM and S. A. HERRES, "Casting and Weldment Stress Relief," *Trans. A.F.S.* **54**, 246 (1946).

20. J. H. SCHAUM, "Stress Relief of Gray Cast Iron," *Trans. A.F.S.* **56**, 265–278 (1948).

1. The part sketched in Fig. 6–19 is a large stainless-steel valve-seat insert, and the surface marked f is to be entirely free of porosity and defects. Assume that the solidification characteristics of the composition to be used are similar to cast steel. However, the alloy is quite susceptible to dross formation in the gate passages.

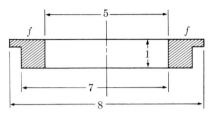

FIGURE 6–19

(a) Calculate a complete mold design which would enable you to produce radiographically sound castings at minimum expense. Furthermore, note that the planned production line calls for a pouring time of four seconds.

(b) In the design you have sketched, indicate the locations at which you would expect hot tears and explain the reasons for each defect briefly.

(c) In machining the outside diameter of the casting shown in Fig. 6–20, the machinist claims that after taking a cut to provide a reference mark and then setting his tool for a given amount of stock removal, he does not obtain the calculated diameter.

(i) Explain the sequence of events which leads to this effect in the casting.

(ii) Assume that after the first cut, the outside diameter is 10.000 in.; the tool is set for what would normally be a 0.200-in. cut, i.e., the spring in the tool holder, etc., is taken into account. Would you expect the new dimension to be greater or less than 9.600?

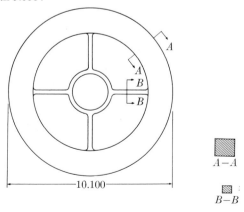

FIGURE 6–20

2. (a) Discuss the mechanism relating sulfur content to the incidence of hot tearing in grade-B cast steel.

(b) Is it possible to avoid hot tearing in a casting of any design at high S-levels, e.g. 0.20%?

3. The casting shown in Fig. 6–21 is poured in steel in a uniform sand mold and is mold cooled. What is the nature of the residual stress at locations A and C? Explain how the condition developed. What would be the stress at B relative to A? At which locations would you expect hot tears? Explain the mechanism.

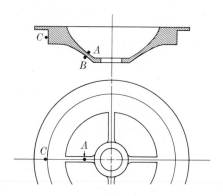

FIGURE 6–21

4. A high-strength (class 50) gray iron was cast to provide a roller-bearing housing (Fig. 6–22). The inside diameter was flame hardened by passing a ring-shaped flame fixture down the bore followed at a 2-in. distance by a water quench. After hardening, a circumferential residual stress of 40,000 psi in compression was encountered uniformly in the bore. Upon tempering at 400°F (204°C) for one hour, this stress decreased to 9,000 psi.

Explain the mechanism underlying stress and development relief. Assume that the level of residual stress in the as-cast condition is negligible. Also explain the occasional occurrence of longitudinal cracks through the hardened layer in the quenched condition.

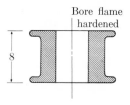

Bore flame
hardened

8

FIGURE 6–22

5. In the wheel sketched in Fig. 6–23, radial cracks are encountered in the flange in the as-cast condition when the phosphorus content of the composition below exceeds 0.07%. The unchilled portion of the wheel is sand cast.

C	Mn	P	S	Si	Fe
1.5	0.60	0.02 to 0.10	0.04	0.50	balance

Explain the presence of the cracks.

In another composition,

C	Mn	P	S	Si	Fe
2.3	0.5	0.06	0.06	1.0	balance

circumferential hot tears were encountered in the hub fillet A, as shown in the figure. What was the cause? Increasing the hub-fillet section at A aggravated the condition. Why? However, the addition of radial fins $\frac{1}{4}$ in. thick and evenly spaced at 2-in. intervals around the fillet, as shown at B, eliminated the condition. Why? In the casting shown, what would be the circumferential stress (tension or compression) on the thread surface at room temperature? Explain the development at B.

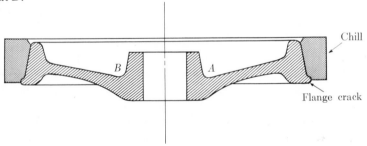

FIGURE 6–23

CHAPTER 7

MOLD PRODUCTION AND PATTERN CONSTRUCTION

7-1 The general problem. To produce castings at a cost which is competitive with other processes, it is essential to achieve economical mold production. Therefore, although in this text the important problems of mold *design* have been discussed first, the practicing engineer should also keep in mind the advantages and the limitations of the molding *processes* that are available. Pattern design and construction are also closely related to the molding process and are discussed at the end of this chapter.

Before we enter into the general discussion, a brief illustration of the close relationship between mold design and the choice of production method may be helpful. As an example, we shall again use the railroad-car wheel introduced in previous chapters.

Cast steel wheels are made successfully by two methods. The first, which employs conventional sand and chiller techniques, was illustrated in Fig. 6-8. The second, which uses permanent molds made of graphite, is shown in Fig. 7-1. In the first case, the chiller at the tread more than offsets the greater mass of the rim, so that the wheel freezes in the order: rim, plate, and hub. A single riser over the hub is therefore adequate to provide metal to compensate for the liquid-to-solid shrinkage of the whole wheel. In the second case, since the entire mold is made of graphite, there is no difference in thermal conductivity of the mold wall. Hence the plate freezes first (between the rim and hub), and it is necessary to feed the rim and hub separately. To accomplish this, four insulated risers are placed around the rim, and the hub is fed from a pressure gating system. Also, to avoid tearing that may be caused by the rigid graphite mold, a nonrestraining, gently curved plate section is used.

Both of the mold designs described are good. The choice depends upon the production method to be adopted. Thus, other designs are found to be particularly suited for other methods of casting, such as die casting or centrifugal casting. With these qualifications in mind, let us review the principles and advantages of the various molding processes.

The main molding processes are green sand, dry sand, core sand, CO_2, shell, investment, permanent mold, and die casting. To illustrate the principles underlying these processes we choose what is perhaps the most interesting approach to the subject, namely a detailed investigation of the widely used method of green-sand casting. Thus we shall analyze the moderately complicated casting of the lever shown in Fig. 7-2, to gain

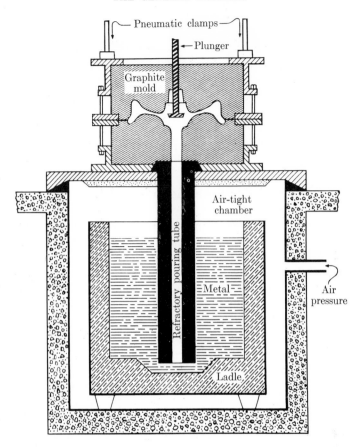

Fig. 7-1. Permanent graphite mold used in the production of steel car wheels. Arrangement of graphite mold, ladle, and tube during pressure-pouring operation. Mold risers are not shown. Pouring speed is closely regulated to prevent mold erosion. Operation is pushbutton controlled. (Amsted Laboratories)

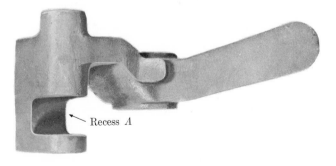

Fig. 7-2. Lever casting (approx. one-third size).

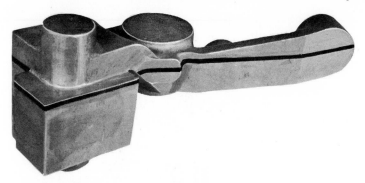

FIG. 7–3. Pattern with parting surface marked (approx one-third size).

some familiarity with the problems of draft and parting line which are encountered in all processes. (Readers unfamiliar with foundry terms will find Appendix I helpful.)

7–2 Green-sand molding. Although the ceramic problems of this material are not discussed until Chapter 8, it should be mentioned here that green sand is a mixture of silica sand, clay, and water (often with special-purpose additives) which can be rammed against a pattern surface to give a firm mold. The pattern is then drawn and liquid metal is poured into the cavity to solidify against the sand surface.

Choice of parting surface. While the lever casting shown in Fig. 7–2 poses a rather simple problem of parting, it does illustrate the effect of casting design upon the choice of a molding method. From an inspection of the casting, it is evident that a two-part mold will be satisfactory. In other words, the two mold halves can be "drawn" or removed from the pattern if the parting surface between the halves follows the line sketched in Fig. 7–3. However, there is a deviation from this line: the recess *A* in Fig. 7–2. To overcome this irregularity, a core is baked separately and inserted in the mold. To permit this insertion, the pattern is enlarged with a "core print," to provide a cavity in the two-part mold. These and other details will be made clearer in the following description.

At this point the proper choice of green-sand molding technique must be made, based upon the number of castings required. For completeness, we shall discuss this problem for four quantity ranges: 1 to 5, 5 to 100, 100 to 1000, and over 1000 castings.

1 to 5 castings. Although we shall discuss molding with a loose pattern, this procedure is not recommended and is only reviewed for the sake of completeness. In any green-sand method, some form of support against which sand can be rammed is needed at the parting surface. For a small number of castings, this support (match) is made of green sand itself, but for large numbers of castings a semipermanent or permanent match is used.

The procedure for making a green-sand match will be described first. A flask is rammed level with green sand, and then the pattern shown in Fig. 7–3 is pressed into the sand. The depth to which the casting is embedded is established by the highest points on the parting surface of the casting, which are pressed level with the top of the sand layer. Care must be taken, also, to establish vertical surfaces properly; if the core print is cocked, it will not draw properly from one side or the other of the mold when the casting is made.

Next, the parting surface is cut as shown in Fig. 7–3. The mechanics of the partition are, to a large extent, self-explanatory, but the diagonal line across the small boss needs some explanation. The placement of this cut is dictated by the angular (not vertical) alignment of the sidewalls of the boss. Note, also, that when there is radius on a sidewall, the parting is at the *end* of the radius adjoining the vertical side of the casting, not at mid-radius or at the end adjoining the top side. After the parting surface has been cut, it is smoothed with a slick or trowel.

It should be emphasized that this part of the operation serves merely to provide a proper match for the pattern. This section cannot be used as part of the final mold, since the sand was merely pressed and not properly rammed against the pattern. To avoid sticking with the drag sand layer in the next operation, a dry parting compound is dusted over the sand.

The flask section is now placed over the match and sand is riddled over the pattern. Backing sand of lower strength and greater permeability than the facing sand is then rammed into place and leveled, a bottom plate is added, and the mold is rolled over. The sand match section is then drawn from the pattern and discarded. An empty cope flask section, which may have been used to hold the match, is placed in position, the downsprue and risers are positioned, the cope section is riddled, rammed, and struck off, and the pouring basin is cut.

To finish the mold, the cope half is drawn from the pattern. At this time the bottom portion of the riser cavities and the riser necks are cut, as well as the runners from the sprue. Now the pattern is drawn, the core is positioned, and the mold is closed and clamped, ready for pouring.

This method has a number of basic disadvantages, both in time consumed and in quality of the castings. Thus, for example, the preparation of the sand match and the many other operations described require about one man-hour per mold, compared with seconds for high-production operations.

5 to 100 castings. The first step in improving the mold for increased production is to make a match of plaster or, at least, a hard baked-sand

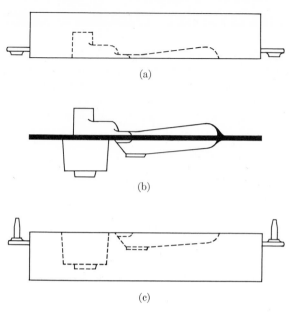

Fig. 7–4. Match plate for lever casting. (a) Cope. (b) Match plate. (c) Drag.

match which can be reused. A portion of the gating or risering system may be attached to the pattern, to eliminate rough sand surfaces as well as variations in gating from mold to mold.

100 to 1000 castings. For sizable quantities of castings, a match plate like that shown in Fig. 7–4 is required. To produce the match plate, a mold is made in the usual manner, and then the cope and drag are separated by a distance of approximately $\frac{1}{2}$ inch. A sand band is rammed in place around the edge of the flask to avoid a runout, and the mold is poured. The resulting casting contains half the pattern on each side of a metal plate, with the proper parting surfaces established and gates and risers in place. All that is now required is to drill the holes in the match plate for the flask pins, and then the cope and drag green-sand molds can be rammed on respective sides of the plate.

A further advantage of this method is that a molding machine (such as a jolt squeeze) may be used instead of the hand-ramming procedure. The drag side is made first. Sand is delivered to the flask from an overhead hopper while the mold is jolted. This process compacts the sand solidly against the pattern surface, but leaves it relatively loose in the remainder of the mold. A squeeze plate is placed inside the flask, and the plate and mold are rolled over together. Sand is now delivered to the cope flask, and then the entire assembly is squeezed by an air ram. The cope and drag are now drawn away from the pattern, the downsprue is cut, and the flask sections are clamped together or weighted, ready for pouring.

Over 1000 castings. For large-volume conveyer operations, separate pattern plates may be made for cope and drag production, and a number of patterns may be mounted on each plate. The use of common risers will also improve the yield. Another high-production method involves mold blowing, i.e. sand is shot into the mold by air pressure and hence compacted. A sand slinger which throws sand from a wheel into the mold cavity is used for larger molds.

7–3 Dry-sand molding. Although more castings are made in green sand than by any other method, this material possesses the inherent disadvantages of high water content and low strength. When water vapor comes into contact with the liquid metal during pouring, there is a strong possibility of gas and dross formation, as will be described in Chapter 9. The low strength can lead to erosion and breakdown of the mold.

The use of a dry-sand mold considerably reduces these hazards. The mold can be made by any of the methods just described, and when the blowing process is used, the pressure can be lower than that required for green sand. After the mold is completed, it must be baked at 300 to 600°F (149 to 316°C) to drive off unbonded water and to intensify the effects of organic binders such as pitch (Chapter 8). In some cases, for economy, only so-called "skin drying" is employed along with surface sprays or washes to develop strength in the surface and to reduce water content. It should be emphasized that at these low temperatures, some but not all of the water bonded to the clay is driven off.

7–4 Core-sand molding. The cores used to form holes or recesses in molds, for example, the core of the lever casting shown in Fig. 7–2, are usually made of oil- or resin-bonded sand. The complete mold consists of cores which are fitted together. In the green condition, such cores are weaker than green sand, but after baking they possess high strength (200 to 400 psi) and are readily handled.

The green core can be rammed by any of the methods discussed for green sand, but for a large-scale production of small-sized cores blowing is the best method. In this case the blow cartridge (Fig. 7–5) is loaded with core sand and the corebox is clamped against the blow plate. Compressed air is then admitted rapidly to the blow chamber, and the sand-air mixture rushes into the corebox. The air leaves either through the fine vent slots at the clamping face of the corebox or through thin vent inserts in the wall of the box. Very elaborate machines of this type have been designed in which retractable inserts in the corebox wall are drawn away from four sides after blowing, thus producing intricate cores. When

FIG. 7–5. Core blowing (lever casting).

the core has no simple flat surface, a drier plate of the proper contour is used as a support until the core is baked.

The baking cycle consists of exposure to 350–500°F (177 to 316°C) for one to six hours. During this period unbonded water is driven off, and the oil oxidizes and polymerizes to form a high-strength bond with the sand. When resin is used instead of oil, the elevated temperature results in polymerization and consequent resinous bonding of the sand.

Since the principal source of strength in dry sand is the oil or resin bond, only enough clay is used to hold the core together in the green condition before baking. Keeping the clay to a minimum is also economical, since clay is a notorious absorber of the expensive oil. When a collapsible core is needed, one uses additives, such as wood flour, which burn out rapidly and help to prevent hot tears in the casting.

7–5 The CO_2 process. In recent years the "CO_2 process" for molding and core making has become increasingly important. Instead of utilizing an oil or a resin that requires heat for bonding, the CO_2 process uses a special sand which is hardened in place by gassing with carbon dioxide. As discussed further in Chapter 8, this special sand contains a solution

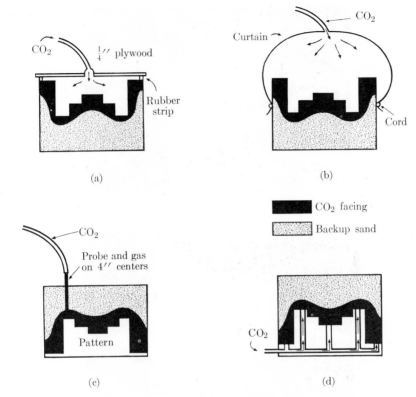

Fig. 7–6. CO$_2$ gassing techniques [2]. (a) Muffleboard. (b) Shower curtain. (c) Lance and probe. (d) Hollow pattern. Vents (not shown) required through entrance and exit surfaces.

of sodium silicate in water. The CO$_2$ gas forms a weak acid that hydrolyzes the sodium silicate, thus forming amorphous silica, which becomes the bond. There is also a bonding action from the sodium silicate itself.

CO$_2$ molding. The sand is mixed with the standard equipment employed for green sand or core sand, the only variation being the use of 2 to 6% of sodium silicate solution. Proprietary compounds are usually added to control the gelling time of the solution. The green strength of the sand is controlled by additions of up to 2% of clay. This sand mix has good flowability and is readily rammed by conventional methods. If desired, the "CO$_2$ sand" may be used as a facing and a cheaper sand employed for backup.

Considerable freedom and exercise of ingenuity are possible in the development of gassing techniques. The details of the gas-liquid reaction are discussed in Chapter 8. The principal mechanics of the process involve uniform flow of gas over the proper period of time and avoidance of channeling. Some of the techniques are illustrated in Fig. 7–6.

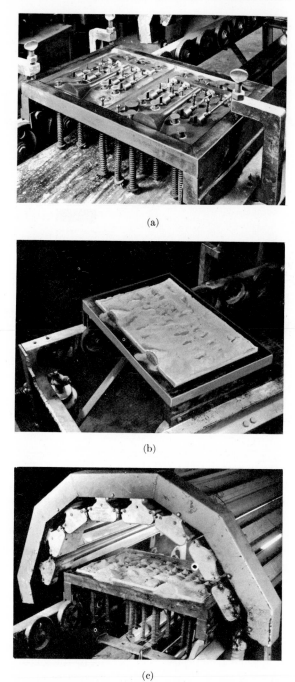

FIG. 7–7. Shell molding technique. (a) Pattern plate. (b) Pattern after investment. (c) Curing mold under heaters. (*Cont.*)

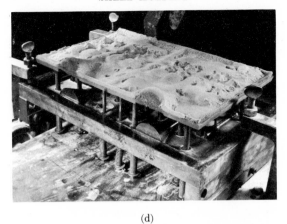

(d)

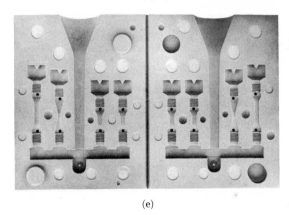

(e)

FIG. 7–7. (*Cont.*) (d) Mold ejection. (e) Finished shell.

CO₂ cores. The CO_2 process is of considerable interest in core making because, with the proper design, the sand can be blown in place and gassed quite rapidly; furthermore, by suitable design, it is also possible to produce hollow cores. One of the great advantages of this method of core making is that, since the core is gassed or "set" in the corebox, no green strength is required. Consequently, one can work with clay-free core sands of high flowability and eliminate the use of driers and the heating cycle.

7–6 Shell molding. Shell molding is particularly adapted to high production rates. In general, better dimensional tolerances and surfaces are obtained than for dry sand, and the process requires only 10 to 20% of the quantity of molding material consumed in dry-sand molding. Although the sand mix is considerably more expensive and the metal pattern equipment is more complicated, the capital investment required to set up a new plant using shell molding is lower than for a dry-sand operation and is competitive with green-sand processes in many cases.

The principles of the process, illustrated in Fig. 7–7, are as follows:

(1) A set of metal pattern plates, similar in many ways to the cope and drag plates for green sand, is heated to 400 to 500°F (204 to 260°C).

(2) Instead of ramming sand against the pattern, a dry mixture of silica sand and thermal-setting resin is dropped upon the pattern, either from a dump box or from overhead louvres. After about 15 seconds of contact, a shell of this mixture, about $\frac{1}{4}$ inch thick, has set against the pattern, and the remaining unaffected powder can be dumped off.

(3) The pattern and the adhering shell are then heated in an oven for less than one minute to cure the shell and restore pattern temperature.

(4) Next, the shell is stripped by means of ejector pins mounted in the pattern. It is then assembled with a mating half from another pattern or a half portion of the same shell to form a complete mold. Unless internal electric heaters are used, the pattern is often preheated further before molding is resumed.

This procedure may be modified in various ways. For example, the cope and drag sections may be mounted on the same pattern plate, with a simple breaker strip for dividing the shell before assembly. Sometimes the entire casting is contained in either a cope or a drag and a flat plate is used to complete the mold, or the casting may even be poured as an open mold. When the casting is symmetrical about the parting plane, it can be poured vertically and the same pattern used for each side.

Sometimes conventional cores are used or shell cores may be prepared. In the latter case, a heated metal corebox is filled with the resin-sand mixture and dumped after the proper interval, and the core is then stripped by separating the pieces of the box. It is not possible to use the customary dry resin-sand mixture for blowing because the mixture would segregate, but sands coated with liquid resin have been developed. Since this process is a fairly recent development, the equipment is still being improved.

One of the critical problems in shell molding is the means of support or backup of the shell during pouring. When the section thickness of the casting is much heavier than that of the shell (for example, in a crankshaft with sections one inch wide), either metal shot, loose sand, or gravel is tamped in place behind the shell. However, when the thickness of the sections is $\frac{1}{2}$ inch or less, the backup may be eliminated, with a consequent reduction in the cost of handling.

Another problem concerns metals which must be poured at high temperatures, such as 0.30% carbon steel. In moderate and heavy sections the metal does not have time to form a solid skin prior to the breakdown of the shell, and the ensuing reaction with the mold material accompanied by rapid formation of iron oxide (a very active flux) produces a rough surface. Stainless steels pose less of a problem because of the lower rate of oxide formation and lower pouring temperatures.

Shell-mold materials. Very rapid changes are taking place in this field, since the resin represents one of the major costs of the shell-molding process. (At an average resin content of 6% and a cost of 30 cents per pound, this item contributes 1.8 cents per pound to the sand cost.) Various liquid resins and other related materials have been or are being developed, and a precoated sand is now widely used. Also, a pure grade of silica sand is required to avoid waste of resin on clay particles or other impurities.

7-7 Investment casting. The outstanding differences between investment casting and the methods just discussed are that (a) liquid material is poured around the pattern, and (b) the pattern is usually used only once. These features result in finer surface detail and greater design flexibility, but may add appreciably to the cost.

The steps in the procedure are illustrated in Fig. 7–8, which shows the production of a cast tensile bar. The risered wax pattern is made in a permanent metal mold by injecting liquid wax. The patterns are then assembled on a gating system, and a soldering iron is used to cement the junctions. To obtain maximum detail the cluster is dipped in precoat slurry, and the precoat is then sprinkled with sand grains to promote a good bond with the body of the investment. Next, the cluster is placed in a cylindrical flask made of heat-resistant tubing, and the liquid investment is poured to fill the flask. About 20 minutes after the assembly has been vibrated to eliminate air bubbles, setting proceeds in the same way as for cement. After the wax has been melted out of the flask in a dewaxing furnace at 400°F (204°C), the mold is fired at 1600° to 1800°F (871 to 982°C) to remove all traces of wax. The mold can either be poured in conventional fashion or can be clamped to a rotating furnace to be poured under pressure or in a protective atmosphere.

Materials. The materials used for expendable patterns may be wax, polystyrene, or frozen mercury. Rubber and other materials have been used for permanent patterns in instances where the pattern can be drawn after investment.

The investment used consists of fine silica bonded by reactions involving tetraethyl silicate or sodium ammonium phosphate. In plaster casting the familiar plaster-of-paris bond is used, but this material is limited to the relatively low pouring temperatures employed for magnesium, aluminum, and a few copper alloys. An important modification of the investment method is the so-called "investment-shell" process, in which the wax pattern is dipped about five times into a ceramic slurry and dried

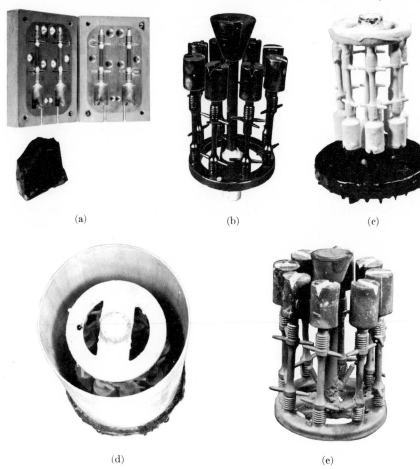

FIG. 7–8. Investment molding technique. (a) Metal mold and wax. (b) Wax cluster. (c) Cluster and dip coat. (d) Cluster in flask. (e) Casting.

between dips. In this way a shell approximately $\frac{1}{4}$ inch thick is built up. The wax is then melted out of the shell or dissolved chemically, and the shell is fired at about 1600°F (871°C) and poured. The advantages of this method over the conventional investment process are: reduced quantity of investment, elimination of flasks, drastically reduced firing time, and simplicity of shake out.

7–8 Permanent-mold centrifugal casting. The permanent mold is another means of achieving low molding costs in quantity production. However, this method gives rise to a number of metallurgical and mechanical problems which tend to restrict its use.

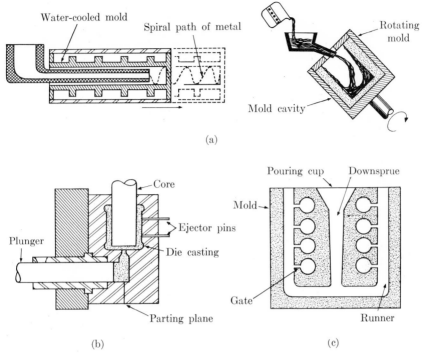

FIG. 7–9. Permanent mold and die casting. (a) Centrifugal castings. (b) Die casting. (c) Gravity casting.

The permanent mold is perhaps most widely employed in the centrifugal casting of pipe. The liquid metal is introduced into a refractory-coated, revolving hollow steel cylinder partially closed at the ends, as shown in Fig. 7–9. After solidification is complete the casting is ejected by a piston. When this method is used, great care must be taken with tear-sensitive metals to ensure free contraction of the metal. In a process developed in Germany, the mold is coated with a thin layer of unbonded silica sand which is held to the walls by the centrifugal action.

Permanent molds are used principally for alloys and designs that are not too susceptible to hot tearing. Thus, for example, the use of metal molds for gray-iron castings has met with success due to the expansion which occurs upon solidification of the eutectic metal molds. Magnesium and aluminum alloys are also gravity cast in permanent molds, particularly when the desired dimensions exceed the capacity of a die-casting machine. Recently, 600-pound steel car wheels have been cast successfully in large graphite molds. (The advantages of a high-conductivity mold in producing narrow-band solidification were discussed in detail in Chapter 2.)

TABLE 7-1

Production method	Patterns (arranged in order of increasing cost)		
Green sand ⎫ Dry sand ⎪ Core sand ⎬ CO_2 ⎭	Loose wood	Mounted wood or plastic	Metal
Shell			Metal
Investment			Metal mold or rubber pattern
Permanent mold			Metal mold
Die casting			Metal mold

7-9 Die casting. Die casting is really a modification of permanent-mold casting in which lower pouring temperatures can be used because the liquid metal is forced into the die under pressure (Fig. 7-9). The lower temperature of the metal, combined with pressure, leads to better surface and dimensional control. The process can be highly mechanized, and intricate dies with many retractable pieces have been used with excellent results. Because of the low pouring temperatures, the process is mainly employed in the production of zinc and aluminum-base alloys, although certain copper-base alloys are also cast in this way.

7-10 Pattern design and construction. Although chronologically the design and construction of a pattern must precede the production of the casting, one cannot tackle these matters until the molding process has been selected, and it for this reason that our discussion of patterns has been postponed until now. We may outline the problem by listing the types of patterns required for the different production methods (Table 7-1) and then discuss similar types as a group.

Pattern for green sand, dry sand, core sand, CO_2. The principles applying to the pattern selection for the first four production methods listed in Table 7-1 are to a large extent identical. The most important factors to be considered are: (1) number of castings to be produced; (2) dimensional tolerances required; (3) type and size of molding machine and flask size; and (4) intricacy of casting shape and size.

Because of the cost of producing an intricate pattern of high accuracy, the number of castings over which the cost may be charged is of great importance. For example, the automotive industry has a demand for large body and fender dies weighing many tons, but usually only one or two of each type are needed. Hence, the cheapest types of wood patterns are used for the dies although all surfaces of the castings are later machined with great care. By contrast, accurately machined metal patterns are employed for crankshaft production, and as a result, a number of surfaces on these parts can be used in the as-cast condition. Thus, when a relatively small number of castings is to be made, under 500 for example, wooden patterns are often specified. Even in this range, however, the choice can vary from a cheap softwood (pine) pattern to a carefully made hardwood (mahogany) type. It is imprudent to resort to a softwood pattern if high quality castings are desired because a soft pattern is rapidly dented. In any event, even for a single casting, the pattern should be mounted on a board with the proper rigging, i.e. gates and risers. Then, if the first trial casting is unsuccessful, there is a definite point of reference from which to review the process. When the gates and risers are cut by hand in the sand, duplication is difficult, and the erosion of sand by the metal is accentuated as well.

For sizeable quantities of castings, metal or plastic (epoxy resin) patterns are made. These materials can usually be machined to closer tolerances, and they are more durable than wood.

The following precautions should be observed in either wood or metal-pattern designs.

(1) The parting line should be chosen so that the smallest portion of the pattern is in the cope. In other words, sand has greater strength in compression, and it is better to form ribs and other details in the drag. The most critical surface should also be placed in the drag because possible defects due to loose sand and inclusions will occur in the cope.

(2) For accuracy and reduced cost, full cores should be used instead of cemented half-cores.

(3) Cores should be minimized by using offset parting where needed.

(4) For volume production, the use of several patterns in a mold and of common risers should be considered.

Patterns for shell molding. For this method metal patterns must be adopted because the process requires heating to 400°F (204°C). (The heating may be accomplished with internal resistance heaters or by external means.) Another design problem is the location of stripping pins to release the shell from the mold. Shell patterns require less draft because the shell is strong at the time of stripping and can withstand a much higher stress level than green sand.

Investment patterns. Pattern design for investment casting is quite different from conventional processes. When expendable wax or polystyrene patterns are to be employed, the first step is to make a master pattern of the casting. From this a mold is made in metal which is then used to produce the expendable patterns. For high production runs involving polystyrene patterns, the mold is usually machined in steel, but for shorter runs utilizing wax patterns, the mold is made of a low-melting metal which is cast around the master patterns. It is obvious that the high accuracy desired in investment castings can be obtained only by means of carefully machined pattern equipment. Furthermore, it is necessary to determine the shrinkage factors of the pattern material, the investment, and the cast metal quite accurately before one can proceed with the construction of the master pattern.

Permanent molds. A permanent mold can either be machined from solid stock or accurately cast to shape. In the latter case, the pattern must be carefully made even though only a few castings are to be produced. In some instances, the accurately cast metal surfaces exhibit longer life than machined surfaces. An important feature of permanent mold construction is the proportioning of mold thickness in order to develop a favorable thermal gradient as the mold is heated during use. To this end, passages for water cooling are often added.

In centrifugal casting, permanent molds are made of graphite as well as of steel. The graphite is more resistant to thermal shock and does not develop the crazed surface often encountered in metal molds.

Die casting. Molds for die casting are machined in almost all cases. Because this process can produce very accurate castings with excellent surface finish, the layout and machining of the molds are carried out with considerable care and are correspondingly expensive.

REFERENCES

1. R. H. S. ROBERTSON and B. EMODI, *Nature*, pp. 153, 339 (1953).
2. R. LEONARD and A. DORFMULLER, JR., "Molding with CO_2," *Foundry,* October, 1957.

GENERAL REFERENCES

Excellent accounts of new molding methods are contained in the following monthly publications: *The Foundry, Modern Castings, British Foundry Trade Journal,* and *Transactions of the British Institute of Foundrymen.*

PROBLEMS

1. Referring to the bearing illustrated in the problem section of Chapter 4, specify the molding methods you would employ in the following conditions:

Material	Quantity
Gray cast iron	1500 total, 5000/day
Aluminum	same
Steel	same

2. Do the same for a typical automotive engine block.
3. Explain the industrial hazards (danger of mechanical injury, smoke, dust) you would anticipate in the different processes.

CHAPTER 8

MOLD MATERIALS; REACTIONS AT
REFRACTORY-METAL INTERFACE

8–1 Introduction. For many years communication between foundry engineers, ceramists, and organic chemists was poor. The introduction of shell molding, which involved advanced organic materials and new investment mixtures requiring complex ceramic bonds, has demonstrated that with proper communication and understanding, tremendously valuable new mold materials can be developed. However, as new metals and alloys (such as the titanium group) are cast, still further advances in mold materials become necessary.

The aim of this chapter is to provide the foundry engineer with a background for understanding and developing new materials, as well as for improving the older ones composed of sand, clay, and water.

It is generally accepted that the properties of a metal casting depend not merely upon the chemical composition, but also upon the internal structure developed by the atomic bonding. In a parallel way, the properties of mold materials, ceramic or organic, also depend upon the types of bonding and the arrangement of atoms. For example, silica, SiO_2, can exist at room temperature as quartz, vitreous silica, and in several other forms. Quartz will shatter upon rapid heating, while vitreous silica, the so-called "fused quartz" of the same composition, can be heated and quenched with impunity. To facilitate the study of the ceramic and organic materials to be discussed, we shall include at this point a review of the nature of ceramic and organic bonds, i.e., the crystal chemistry of these materials. Thus, we shall describe the four types of bonding, ionic, covalent, molecular, and metallic, and then pass on to the principles governing the development of ceramic and organic structures from simple units such as the SiO_4^{4-} tetrahedron. As examples of ceramic materials, we shall discuss the structure of various types of silica, of clays, and of miscellaneous silicates such as olivine and zircon. We shall then go on to the clay-water and hydraulic bonds present, say in plaster and hydrolyzed ethyl silicate. For the organic materials, we propose to describe the polymerization and cross linking of phenol or urea formaldehyde resins (used in shell molds) as well as the oxidation of core oils.

Having reviewed this basic background material, we shall interpret the engineering behavior of green sand, core sand, sodium silicate (CO_2-process) sand, and shell and investment materials in conventional tests.

The reaction between metal and refractory, which is really the final test of the merit of any mold material, is discussed in the concluding portion of this chapter.

8–2 Review of crystal structures and bonding forces.* A study of crystal chemistry reveals that there are four principal types of bonding, as shown in Fig. 8–1.

Ionic bonding involves the complete transfer of one or more electrons from one element to another, resulting in more stable electronic configurations for both. In the formation of sodium chloride, for example, the sodium atom gives up one electron and thus becomes a positively charged ion, and the chlorine atom accepts this electron, to become a negatively charged ion. These ions, which attract each other strongly, combine to form the crystal lattice of salt. The spacing between atoms reaches equilibrium only when the electrostatic attractive forces resulting from ionization are balanced by the ion-core (nucleus) repulsive forces which do not operate unless the atoms are in fairly close proximity. Hence the ionic bond is very strong.

There is often some confusion about the mechanism which holds the NaCl *crystal* together. Although each Na$^+$ has given up an electron to *some one* Cl$^-$, the electrostatic attraction holding the lattice together acts between *all* Cl$^-$ ions and Na$^+$ ions. As a result, each Cl$^-$ is surrounded by six equidistant Na$^+$ ions, and vice versa. For convenience, we can consider that the bond holding together each ion pair is one-sixth of the total bond. However, some attraction is exerted by more distant ions, and this adds to the binding energy. The ionic radius is defined as the radius of the spheres (see Fig. 8–1), and is calculated from x-ray diffraction data. Typical values, expressed in angstroms (10^{-8} cm), are given in Table 8–1.

In *covalent bonding*, which is the strongest type of bond, two or more atoms share valence electrons. The resulting structure is more stable and of lower energy than that of the individual atoms. For example, as two atoms of chlorine approach each other closely, each shares one of its electrons with the other, and thus an electronically stable Cl$_2$ configuration containing filled outer shells is formed. Similarly, in diamond and in organic materials, carbon forms covalent bonds.

Intermolecular, or *Van der Waals, bonding* is the weakest type, since only residual electronic, gravitational, and other attractive forces not completely satisfied by other bonds are involved. For example, in the molecule CH$_3$Cl (methyl chloride) four covalent bonds exist between the

---

* It is assumed that the reader is familiar with the usual valency concepts of general chemistry. If not, a review of any college-level text will be helpful.

TABLE 8–1 [14]

ATOMIC WEIGHTS AND IONIC RADII

	Symbol	Atomic number	Atomic weight*	Ionic radius† CN 6
Aluminum	Al	13	26.97	0.57
Antimony	Sb	51	121.76	0.62 (Sb^{5+}), 0.90 (Sb^{3+})
Argon	A	18	39.944	
Arsenic	As	33	74.91	0.40 (As^{5+}), 0.69 (As^{3+})
Barium	Ba	56	137.36	1.36
Beryllium	Be	4	9.02	0.34
Bismuth	Bi	83	209.00	0.74 (Bi^{5+}), 1.20 (Bi^{3+})
Boron	B	5	10.82	0.22
Bromine	Br	35	79.916	1.96
Cadmium	Cd	48	112.41	1.00
Calcium	Ca	20	40.08	1.01
Carbon	C	6	12.010	0.18
Cerium	Ce	58	140.13	1.18
Cesium	Cs	55	132.91	1.67
Chlorine	Cl	17	35.457	1.81
Chromium	Cr	24	52.01	0.65 (Cr^{3+})
Cobalt	Co	27	58.94	0.77 (Co^{2+}), 0.65 (Co^{3+})
Columbium	Cb	41	92.91	0.70
Copper	Cu	29	63.57	0.96 (Cu^{1+})
Dysprosium	Dy	66	162.46	
Erbium	Er	68	167.2	1.04
Europium	Eu	63	152.0	1.13
Fluorine	F	9	19.00	1.33
Gadolinium	Gd	64	156.9	1.11
Gallium	Ga	31	69.72	0.62
Germanium	Ge	32	72.60	0.48
Gold	Au	79	197.2	1.37
Hafnium	Hf	72	178.6	0.84
Helium	He	2	4.003	
Holmium	Ho	67	164.94	1.05
Hydrogen	H	1	1.0080	
Indium	In	49	114.76	0.87
Iodine	I	53	126.92	2.19 (I^{-1})
Iridium	Ir	77	193.1	
Iron	Fe	26	55.85	0.79 (Fe^{2+}), 0.67 (Fe^{3+})
Krypton	Kr	36	83.7	
Lanthanum	La	57	138.92	1.14
Lead	Pb	82	207.21	1.32
Lithium	Li	3	6.940	0.69
Lutecium	Lu	71	174.99	0.99
Magnesium	Mg	12	24.32	0.75
Manganese	Mn	25	54.93	0.91 (Mn^{2+}), 0.52 (Mn^{4+})
Mercury	Hg	80	200.61	1.11
Molybdenum	Mo	42	95.95	0.67 (Mo^{4+})

TABLE 8–1 (*cont.*)

	Symbol	Atomic number	Atomic weight*	Ionic radius† CN 6
Neodymium	Nd	60	144.27	1.08
Neon	Ne	10	20.183	
Nickel	Ni	28	58.69	0.74
Nitrogen	N	7	14.008	
Osmium	Os	76	190.2	
Oxygen	O	8	16.0000	1.32
Palladium	Pd	46	106.7	
Phosphorus	P	15	30.98	0.34
Platinum	Pt	78	195.23	
Potassium	K	19	39.096	1.33
Praseodymium	Pr	59	140.92	1.16
Protactinium	Pa	91	231	
Radium	Ra	88	226.05	
Radon	Rn	86	222	
Rhenium	Re	75	186.31	
Rhodium	Rh	45	102.91	0.69
Rubidium	Rb	37	85.48	1.48
Ruthenium	Ru	44	101.7	
Samarium	Sm	62	150.43	
Scandium	Sc	21	45.10	0.81
Selenium	Se	34	78.96	1.95 (Se^{-2})
Silicon	Si	14	28.06	0.41
Silver	Ag	47	107.880	1.20
Sodium	Na	11	22.997	0.98
Strontium	Sr	38	87.63	1.18
Sulfur	S	16	32.06	1.81 (S^{-2})
Tantalum	Ta	73	180.88	0.68
Tellurium	Te	52	127.61	0.85
Terbium	Tb	65	159.2	
Thallium	Tl	81	204.39	1.47
Thorium	Th	90	232.12	1.06
Thulium	Tm	69	169.4	
Tin	Sn	50	118.70	0.73 (Sn^{4+})
Titanium	Ti	22	47.90	0.65 (Ti^{4+})
Tungsten	W	74	183.92	0.67
Uranium	U	92	238.07	1.01 (U^{4+})
Vanadium	V	23	50.95	0.59
Xenon	Xe	54	131.3	
Ytterbium	Yb	70	173.04	
Yttrium	Y	39	88.92	0.87
Zinc	Zn	30	65.38	0.79
Zirconium	Zr	40	91.22	0.82

* International atomic weights, 1941.
† Ionic radii taken mainly from Stillwell (1938) and averaged with other values, for coordination number 6.

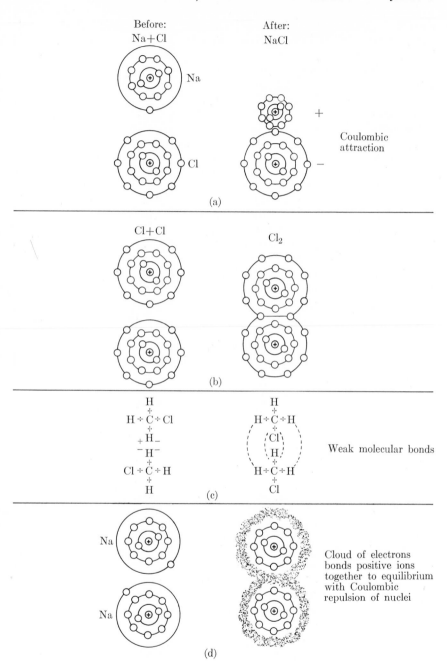

FIG. 8–1. Different types of bonding. (a) Ionic. (b) Covalent. (c) Molecular. (d) Metallic.

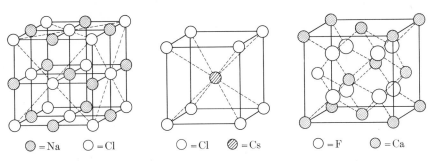

$= Na$ $= Cl$ $= Cl$ $= Cs$ $= F$ $= Ca$

FIG. 8–2. Sodium, cesium chloride, and calcium fluoride structures.

carbon and the four other atoms. There is, however, a small polarity or asymmetry of the electron cloud of the molecule that permits attraction of another molecule. These intermolecular forces are very weak, as evidenced by the fact that methyl chloride is liquid at the ambient temperature. The low boiling point, i.e., the point at which the intermolecular bonds are broken to form a vapor, is a further indication of weakness. It should be emphasized that the *covalent* bond is, of course, not broken by vaporization. The molecules of CH_3Cl still exist and are quite stable in the gaseous phase.

In *metallic bonding*, which is relatively strong, metal atoms give up outer-shell electrons to a common electron "gas" which acts to bond the positive metal ions in a metal crystal. Hence a given valence electron is not identified with a given atom or atoms. Examples of this type of bonding are found in all the metals and alloys. The relatively free movement of electrons produces the well-known high electrical conductivity of metals.

To understand more complex structures such as those exhibited by silica or kaolinite (clay), $Al_2Si_2O_5(OH)_4$, we must add to the valence concept the consideration of ionic packing. In other words, given an electrically balanced assembly of ions, how will these ions pack in a stable crystal? This question is very simply answered by comparing the NaCl and CsCl structures (Fig. 8–2). The positively charged Na and Cs ions have the same valence, yet an examination of the figure shows that in NaCl each Na^+ has six immediate neighbors, while in CsCl each Cs^+ has eight. The number of nearest (equidistant) neighbors is called the *coordination number* (CN). A simple geometrical calculation shows that if the ions are considered as spheres, it is not possible for eight chlorine ions of radius 1.81 A to touch a single Na^+ ion (radius 0.98 A), while they can come into contact with the Cs^+ ion (radius 1.67 A). [Draw a (110)-plane at 45° through the CsCl structure and show that the smallest sphere which will touch all four Cl ions is of radius 1.11 A].

TABLE 8–2

Maximum CN	Radius$_{(cation)}$/radius$_{(anion)}$
1	—
2	0.155
3	0.155–0.225
4	0.225–0.414
6	0.414–0.732
8	0.732–1
12	1.0

The radius ratios (radius$_{[cation\ (+)]}$/radius$_{[anion\ (-)]}$) and the corresponding coordination numbers are listed in Table 8–2. It should be noted that it is possible to have various coordination numbers whether the cation and anion have the same or different magnitudes of charge. In the CsCl structure, one-eighth of each Cs^+ valence per surrounding Cl^- ion may be thought of as shared; in the NaCl, one-sixth of each Na^+ valence will be similarly shared. A count of ions throughout the lattice will show that the stoichiometric ratios of one Na^+ or Cs^+ to one Cl^- are obeyed. In the CaF_2 structure, each Ca^{2+} valence bond is shared with four surrounding F^- ions (Fig. 8–2).

Another consideration in developing structures is that one element may replace another element of similar radius ($\pm 15\%$) and valence without changing the crystal structure. For example, from Table 8–1 we can see that the radius of iron (Fe^{2+}) is within 15% of the ionic radius of magnesium, and iron could therefore replace magnesium in the MgO lattice. Potassium, on the other hand, would not be expected to replace sodium in appreciable amounts in the NaCl lattice. Both of these inferences are confirmed by the existence of mineral deposits which encompass a series of minerals from fosterite (Mg_2SiO_4) to fayalite (Fe_2SiO_4); however, there is no replacement of sodium by potassium in the mineral halite (NaCl), although potassium salts are found in the same beds.

Let us now review the crystal structures of some of the more important ceramic materials used in molds.

8–3 Silica and silicate structures. Silica and the silicates (as in clay) represent the most important group of mold materials. The structures can be readily understood by means of a review of the silicon-oxygen bond. Since the radius ratio $Si/O = 0.41/1.32 = 0.31$, the coordination number is 4. The silicates as well as the different types of silica (SiO_2) are therefore constructed from a single group, the SiO_4^{4-} tetrahedron (Fig. 8–3).

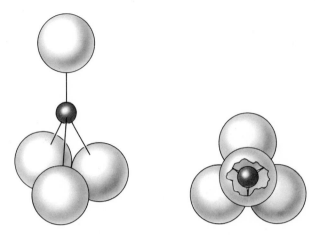

Fig. 8–3. The SiO_4^{-4} tetrahedron.

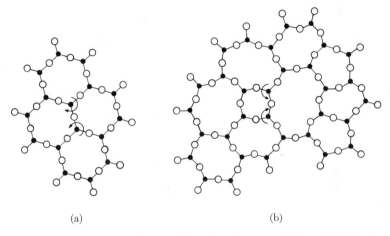

(a)   (b)

Fig. 8–4. Silica structures in two-dimensional representation. (a) Quartz. (b) Vitreous silica.

The silicon-oxygen bond is partly covalent and partly ionic, each oxygen sharing two electrons with a given silicon. Four silicon electrons and four electrons from the surrounding oxygen ions satisfy the silicon bonds. To have a stable outer ring of eight electrons, each oxygen ion requires an additional electron, either shared or from an ionic bond with a metal. Therefore, in the silicates we find a combination of covalent and ionic bonds. The metallic element satisfies the requirement of an additional oxygen bond by donating an electron. The end point in the silicate series is silica itself, in which each oxygen atom is attached to two silicon atoms, as shown in Fig. 8–4(a). This structure represents the most common form of silica, quartz.

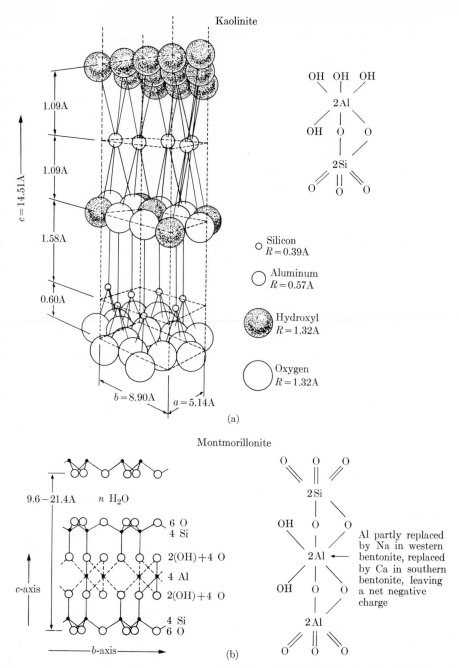

FIG. 8–5. Sheet structure of clay minerals. (a) Kaolinite $(OH)_4Al_2Si_2O_5$, perspective exploded in c-direction. (b) Montmorillonite $(OH)_4Al_4Si_8O_{20}\cdot nH_2O$.

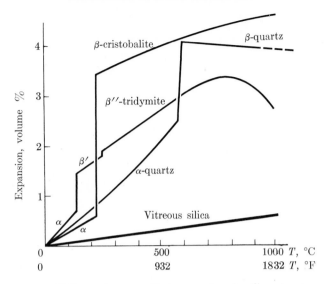

FIG. 8–6. Volume changes with temperature in silica structures.

By contrast, in the clay materials the silicate tetrahedra are linked to one another in only two dimensions, forming a sheet with foreign atom bonds in the third dimension (Fig. 8–5). Here we may pause to summarize our experiences with quartz and clay. The quartz structure has covalent bonds and equal strength in all directions. There is no preferential cleavage, and the covalent bonds lead to high hardness. On the other hand, the structure of clay readily lends itself to cleavage parallel to the sheet of strong covalent bonds. The cleavage is through the weak Van der Waals' bonds between layers, as we shall describe later.

Silica is commonly encountered in four different structures, all built from the SiO_4 tetrahedron. The three crystal structures, quartz, tridymite, and cristobalite, differ in the orientation and the density of the packing of the tetrahedra. In addition, vitreous silica, which persists for very long periods of time, must also be considered. The equilibrium temperature ranges for these four modifications are:

Quartz	up to 870°C (1598°F),
Tridymite	870 − 1470°C (1598 − 2678°F),
Cristobalite	1470 − 1710°C (2678 − 3110°F),
Vitreous silica	above 1710°C (3110°F).

It should be clearly understood that transformation from one of the above types to another is extremely sluggish. For example, the transformation rate of vitreous silica to quartz at room temperature is very slow because the thermal activation is insufficient to accomplish the necessary extensive rearrangement of bonds. During heating and cooling, each of the above structures undergoes various changes which do not require bond rearrangement and which are best understood by considering the changes in volume with temperature (Fig. 8–6). The volume change in quartz which is pronounced compared with that in vitreous silica explains the use of the latter material in thermocouple protection tubes or in special molds where cracking must be avoided. It is important to pour investment molds made of quartz sand at temperatures above the inversion point ($\alpha - \beta$ quartz), not only to fill delicate details but to avoid mold shattering.

The diagrams of the crystal structures of quartz and vitreous silica (Fig. 8–4) provide an explanation of the behavior described above. The vitreous silica lattice is apparently more loosely bonded than the rigid quartz crystal and undergoes only slight volume changes with changing temperature.

8–4 Clays. The foundry engineer is usually familiar with the fact that there are different types of clay, such as kaolinite and montmorillonite (bentonite). However, he is often unable to explain in basic terms the behavior of these materials at elevated temperatures.

Let us refer again to the sheet structure formed when SiO_4 tetrahedra are joined to one another in one plane. One then has a layer of oxygen ions with a single free bond per oxygen ion on *one* side of the plane. The three other oxygen ions per SiO_4 tetrahedron have their bonds satisfied by linkage with oxygen ions from an adjacent tetrahedron. If these excess oxygen bonds are satisfied by a layer of aluminum ions, the structure will be that of kaolinite (Fig. 8–5). The additional aluminum bonds are satisfied by OH^- ions. Since these OH^- ions have approximately the same radius as oxygen ions, the OH^- ions and the oxygen ions together form a layer. To accommodate all the OH^- ions, still another layer, above the aluminum ions, is formed. The bonding action of kaolinite in molding sands results from the free ionic bonds at the edges of a crystal. These broken bonds are capable of adsorbing foreign ions such as Na^+, H^+, etc. A clay's maximum capacity to adsorb ions is called its *base exchange** and is expressed in milliequivalents per 100 grams of clay. (A milliequivalent = atomic weight in grams/1000; e.g., 0.23 gm for Na^+.)

---

* The measurement of the base exchange is conducted in a special electrodializing cell in which all adsorbed ions are stripped off and replaced by H^+ ions. The resulting compound is called a "hydrogen clay" which is then titrated with a basic solution. The solution is buffered by the replaceable hydrogen ions of the clay, and when these have been neutralized, a pH of 7 is reached rapidly. The amount of base required to neutralize the hydrogen clay represents the base-exchange capacity.

The structures of sodium and calcium montmorillonites, the principal minerals in western and southern bentonite, respectively, give rise to a much greater base-exchange capacity. Unlike the kaolinite structure which contains one layer each of aluminum and silicon atoms, these structures have silica tetrahedra on *both* sides of the aluminum ions (Fig. 8–5). The silicon ions, being at the center of SiO_4 tetrahedra, have coordination number 4, while the aluminum ions have coordination number 6, since they are at the center of four oxygen ions and 2 OH^- ions. (The O^{2-} and OH^- have very nearly the same radius.) In the latter case

$$\frac{R_{Al=0.57}}{R_{O,OH=1.32}} = 0.432,$$

which is in the correct range for $CN = 6(0.41 - 0.73)$. As indicated in the figure, some of the Al^{3+} ions may be replaced by Mg^{2+} in western bentonite. This is just about reasonable from the standpoint of the ionic diameters for Mg^{2+} and for Al^{3+}, which are 0.75 A and 0.57 A, respectively. However, a net negative charge results because aluminum is replaced by magnesium on an ion-for-ion basis. We may verify this observation by the following typical analysis of Black Hills bentonite:

$$(Al^{3+}_{1.67}Mg^{2+}_{0.33})Si^{4+}_4O^{2-}_{10}(OH^-)_2,$$

yielding a net charge of -0.33 which, however, is neutralized by loosely held or exchangeable Na^+ ions, so that we obtain the formula:

$$(Al^{3+}_{1.67}Mg^{2+}_{0.33}Na^+_{0.33})Si^{4+}_4O^{2-}_{10}(OH^-)_2.$$

(The Na^+ ion is loosely held, because it will not fit in the existing space lattice.) The above analysis is representative of the so-called sodium bentonites encountered in the West. As would be expected, the pH is on the alkaline side (approximately 9) because some solution of Na^+ takes place in water.

The southern or "calcium" bentonites differ from sodium bentonites in that iron, Fe^{2+}, rather than magnesium replaces the aluminum. Since in this case, the exchangeable ion is principally calcium, with some hydrogen instead of the sodium of western bentonite, the pH is acid. It is interesting to note that many of the German bentonites are of the "calcium" montmorillonite type and are converted to the sodium type by chemical processing involving ion exchange. (If the clay is treated with a solution containing a high concentration of Na^+, the Ca^{2+} in the bentonite is replaced with Na^+ and precipitates as a calcium salt.) The base-exchange value for bentonite is many times that for kaolinite, whose exchange properties depend primarily upon *fractured* bonds at the edge of the layer.

There is apparently a relationship between the charge deficiency in the central layer of montmorillonite and the ability to adsorb large quantities of water between the adjacent clay platelets. Depending upon the water content between SiO_4^{4-} layers (Fig. 8–5), the distance between the sheets of aluminum ions may vary from 9.6 to 21.4 A in the sodium montmorillonite.

8–5 The clay-water-silica bond. In the discussion of bonding it was pointed out that clay particles possess a surface charge either because of rupture of the bonds (kaolinite) or because of lack of electrical neutrality (montmorillonite). In both cases, exchangeable ions such as Na^+ may be adsorbed to provide neutrality. As a result, water molecules are attracted either between the clay plates, as in sodium montmorillonite, or at the edge, as in kaolinite. The water molecules are polar, and as the first layer adjusts to the electric field of the clay, a second water layer is attracted to the first. The attractive force of the second layer is less than that of the first, and thus succeeding layers are only weakly attracted. Surface attraction also exists between water and quartz because of an unequal balance of charges at the fractured quartz surface. A linkage, quartz-water-clay-water-quartz (or clay), is therefore set up throughout the molding sand.

One of the important considerations in developing the proper distribution of clay is the pH of the sand. Figure 8–7 indicates that under acid

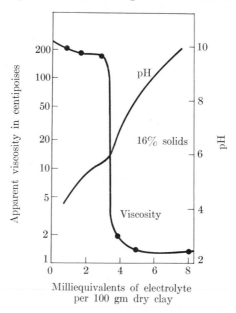

FIG. 8–7. Effect of pH on flocculation of clay. (After Norton [1])

conditions (low pH) the viscosity is high due to flocculation (flocking together or balling) of the clay platelets. As the pH changes from 5 to 7, viscosity changes from 200 to 2. For optimum distribution of clay in a sand mixture, slightly basic conditions are therefore desirable. One explanation of this mechanism is that at low pH the excess H^+ ions are bonded to the clay particles, while under basic conditions Na^+ or other ions are bonded to the clay. The attractive forces between the clay particles are apparently much higher in the former case, thus leading to flocculation. This phenomenon may be due to the differences in ionic radii of sodium and hydrogen, which result in different potential fields around the clay particles.

8–6 Hydraulic bonds. In addition to the clay-water type of bond, a much stronger, so-called hydraulic bond can be developed. Although the term "hydraulic" implies the participation of water, the binding agent is, in fact, a new material crystallized to form a coherent skeleton. The water may participate as water of crystallization in the new structure. The decomposition of ethyl silicate resulting in a coherent silica structure is one example.

Under proper conditions of pH and catalysis, ethyl silicate decomposes to form ethyl alcohol and silica. In preparing an investment mold, a slurry consisting of activated ethyl silicate, as the bonding material, and silica sand is poured around expendable wax patterns. As hydrolysis takes place, a network of silica is gelled around the sand grains, thus providing a bond. Upon heating, surplus water is eliminated from the gel, and the bond is strengthened. The structural equation is

$$
\begin{array}{c}
\\
-\overset{\displaystyle |}{\underset{\displaystyle |}{Si}}- \\
O \\
\end{array}
$$

$$
\underset{\text{(ethyl silicate and water)}}{
C_2H_5-\overset{\displaystyle C_2H_5}{\underset{\displaystyle C_2H_5}{Si}}O_4-C_2H_5 + 2H_2O \rightarrow}
\quad
-\overset{|}{Si}-O-\overset{\displaystyle |}{\underset{\displaystyle |}{Si}}-O-\overset{|}{Si}- + \underset{\text{(ethyl alcohol)}}{C_2H_5OH}
$$

$$
\underset{\text{(amorphous silica bond)}}{-\overset{|}{\underset{|}{Si}}-}
$$

A similar reaction that occurs with hydrated sodium silicate is of importance in the CO_2 sand process previously discussed. A core is made with a solution of sodium silicate well distributed among the sand grains. As CO_2 gas passes through the core it forms a weak acid, H_2CO_3, in the sodium silicate solution which hydrolizes the silicate to form a silica gel, as in the case of the ethyl silicate. One drawback of this procedure is the presence of a sodium salt residue which can act as a flux to lower the refractoriness of the sand mixture. However, this can be counteracted by the use of a suitable mold wash.

Another hydraulic bond of practical importance is that arising in the hydration of plaster of paris to form gypsum. Several structures have been reported in the $CaSO_4$ system: $CaSO_4 \cdot 2H_2O$ (gypsum), $CaSO_4 \cdot \frac{1}{2}H_2O$ (plaster of paris), and $CaSO_4$ (anhydrite). When gypsum, a naturally occurring mineral, is heated, some water is driven off, leaving $CaSO_4 \cdot \frac{1}{2}H_2O$ (plaster of paris), and the gypsum structure is destroyed. When the plaster of paris is again mixed with water, growth of a coherent network of new gypsum crystals takes place.

It is interesting to compare the gypsum structure with those of silicates. The basic unit of gypsum is the SO_4^{2-} tetrahedron, whereas SiO_4^{4-} is the basic unit for the silicates. These units are bonded to calcium to form a layer structure:

$$
\begin{array}{c}
| \\
Ca \\
| \\
-Ca-SO_4-Ca- \\
| \\
Ca \\
|
\end{array}
$$

Sheets of H_2O molecules are interleaved with the calcium sulfate sheets, as in clay, and cleavage occurs perfectly parallel to the sheets. The water molecules are bonded weakly, and hence the gypsum is converted to $CaSO_4 \cdot \frac{1}{2}H_2O$ rapidly at 190 to 200°F (88 to 93°C). Since conversion to anhydrous calcium sulfate requires a much higher temperature, it follows that there are stronger bonding forces with the water in the hemihydrate than in gypsum.

Complex phosphate bonds are also used in a variety of investment molds. The development of a new bonding structure is brought about by the interaction of monobasic ammonium and sodium phosphate and magnesia in water of low (acid) pH.

In the majority of the above cases, the formation of the gel or new structure is accompanied either by an expansion, or by a rather negligible decrease in volume. As a result, the pattern is reproduced in true detail.

8–7 Other silicates. In addition to silica and clay, olivine and zircon are occasionally used as mold materials.

Olivine is a mineral of varying chemistry in which Fe^{2+} is substituted in solid solution for part of the Mg^{2+} in Mg_2SiO_4. The structure, there-

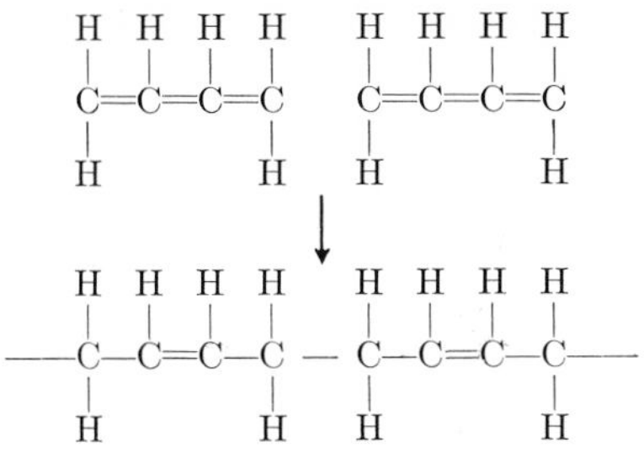

(a)

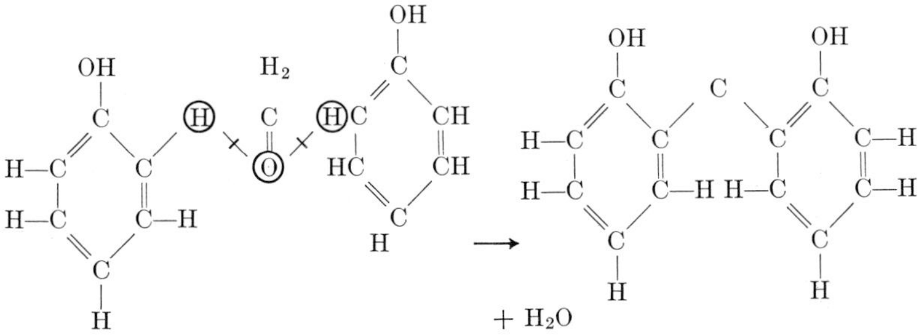

(b)

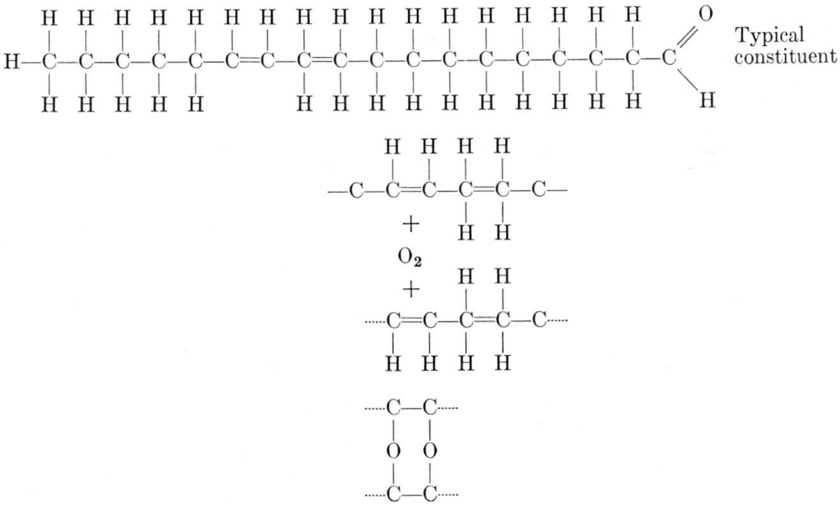

(c)

Fig. 8–8. (a) Addition polymerization. (b) Condensation polymerization. (c) Cross linking.

fore, consists of SiO_4^{4-} tetrahedra bonded together by Mg^{2+} or Fe^{2+}. Since the radius ratios are

$$\frac{Mg^{2+}}{O^{2-}} = \frac{0.75}{1.32} = 0.57 \quad \text{and} \quad \frac{Fe^{2+}}{O^{2-}} = \frac{0.79}{1.32} = 0.59,$$

the coordination number is 6 in either case. The structure, therefore, consists of SiO_4^{4-} islands, each oxygen-bonded to one silicon and three magnesium atoms.

Zircon is $ZrSiO_4$. Since the ionic radius of zirconium is 0.82 A, a coordination number of 6 would be expected (radius ratio = 0.82/1.32 = 0.67). The structure forms a very tightly bonded, high melting-point refractory.

8–8 Organic bonds. In a number of molding materials, such as core sand and shell sand, organic bonds are used in addition to, or in place of, ceramic types. In general, the types of organic bonding employed are polymerization and cross linking.

Polymerization can be subdivided into two principal kinds: addition and condensation. A classic example of addition polymerization is the conversion of butadiene to rubber which involves the breaking of a double bond in each butadiene molecule to link the two molecules (Fig. 8–8a). Giant polymer molecules of increasing strength and melting point continue to form in proportion to the degree of polymerization. In condensation polymerization (Fig. 8–8b), two different types of molecules react to give the polymer *plus* a condensation product such as water or alcohol. Thus, for example, water is the product in the phenolformaldehyde or ureaformaldehyde reaction used in shell sands.

Cross linking (Fig. 8–8c) involves the joining of two molecules at double-bond points by a foreign atom such as oxygen in air-setting linseed oil. The resulting large, more complex molecule is stronger and has a higher melting point than the individual molecules.

THE BEHAVIOR OF VARIOUS MOLD MATERIALS

8–9 Green sand. In the light of the preceding discussion, we shall now review the behavior of the various materials used in the molding processes described in Chapter 7. Because of the importance of green sand in the foundry industry, we shall begin our review by discussing in detail the types of quartz sand available, grain size and distribution, various characteristics determining the molding behavior, and a number of acceptance tests, devised for the purpose of evaluating the molding properties of a

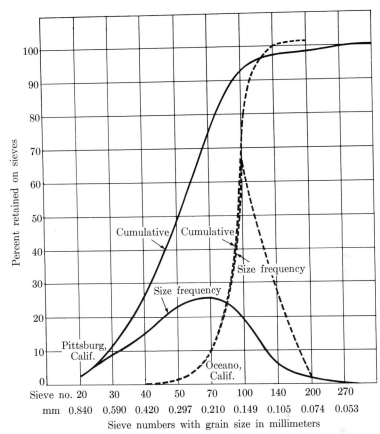

FIG. 8–9. Grain distribution in different sands [15].

material. Furthermore, since in green-sand molding, which involves ramming a mold of "moist" or "green" sand, the quartz-clay-water bond is the most important consideration, we shall also discuss the clays that are used commercially and the extent to which water content contributes to the final mold properties. The part played by such minor constituents as sea coal, pitch, and dextrin will also be considered.

As would be expected from the structural considerations presented in preceding sections, quartz grains show no pronounced cleavage and therefore are rounded or semirounded. The roundest grains are obtained from deposits in areas where the sand has been subjected to wind action over long periods of time (bank sand), as in the Ottawa Sands. Lake or dune sand is of medium angularity, and in areas where the sand has been subjected to heat shock or crushing, the grains are of maximum angularity. Angularity increases with use.

Grain size distribution. Grain size and grain distribution are important factors in the selection of sands for the various processes. These physical characteristics are measured by means of screen tests: a weighed representative sample of the sand is sifted through screens with standardized holes and the percentage of the material that is retained on each of the successively finer sieves is determined. The results of such tests are expressed in two ways: as graphs of cumulative percent retained, or size frequency (Fig. 8–9), or as a single American Foundrymens' Society grain-size number, which is a weighted average. The A.F.S. number, however, is satisfactory only for comparison of two sands of similar size distribution. If a sand exhibits a wide variation in size (such as the Pittsburgh sand whose size distribution is compared with that of Oceano sand in Fig. 8–9), the properties indicated on a graph will be quite different from those implied by the A.F.S. number. For example, permeability, which is a function of the void space, will be lower for a sand distributed over many screens than for one retained by only three or four screens. Also, if a mixture of two different sands is used in a foundry, a double peak not indicated by the A.F.S. number will appear on the distribution curve.

Acceptance tests for green sand. A series of test procedures has been developed for the purpose of evaluating the molding behavior of sands. The principal tests are designed to measure the green compressive strength and deformation, the permeability, and the moisture content. Full details are given in the *Foundry Sand Handbook* (A.F.S. 1952).

(1) Green compressive strength is determined by ramming a cylinder 2 inches in diameter and 2 inches long under standard conditions and then compressing it under increasing load. Typical results [11] for various sand-clay-water combinations are given in Fig. 8–10. In each case, there is a critical water content for maximum strength, which probably corresponds to the point at which sufficient water is present to form the greatest number of clay-water-quartz bonds. Beyond this point, the excess water causes a weak link in the bond and lowers the strength.

Greatest strengthening per percent of clay added is obtained with the bentonites, probably because of the finer dispersion and the differences in bonding. Variation in strength with water content is greater for the bentonite-bonded sands; therefore moisture control is more critical in these materials.

In more recent work [7], the effects of silica flour fines have been determined. These may tie up clay to an extent of 1 to 1.5% bentonite per 1% flour and lower the strength considerably.

(2) Deformation during the green-compression test is receiving increasing attention, inasmuch as it gives evidence of the "toughness" of a sand. Deformation is measured by means of a dial gage attached to

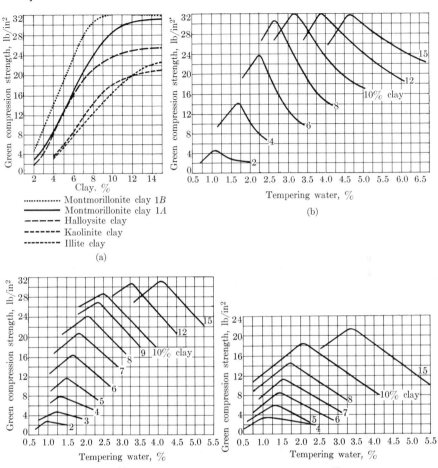

FIG. 8–10. Effect of variations in clay and water content on the strength of sand mixtures [11]. (a) Maximum strength of different clay-sand mixtures. (b) Effect of water content on strength of southern bentonite-sand mixtures. (c) Effect of water content upon green strength of western bentonite-sand mixtures. (d) Effect of water content upon strength of kaolinite-sand mixtures.

the green-sand testing head. A linear increase in deformation with increasing moisture is obtained for both natural and synthetic sands. This is one of the reasons why molders prefer to use a sand with a high moisture content rather than one whose moisture content is characteristic of maximum strength.

(3) Permeability is determined by measuring the rate of air flow through the compression specimen discussed above. A simpler method uses the pressure of the specimen against a calibrated orifice.

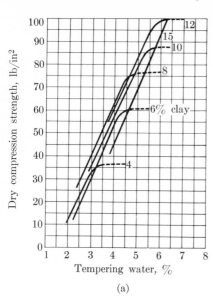

(a)

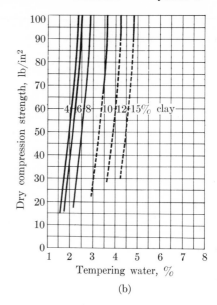

(b)

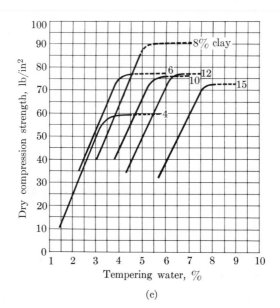

(c)

FIG. 8–11. Effect of initial water content on dry strength. (a) Southern bentonite clay. (b) Western bentonite clay. (c) Kaolinite clay.

(4) Moisture in green sand is measured by drying a weighed sample under a hot air blast and then reweighing. This test indicates only the excess (unbonded) water.

(5) Flowability is a property of sand that is difficult to define or to measure. It is the ability of a sand to behave like a fluid, i.e., when rammed, to flow to all portions of a mold and distribute the ramming pressure evenly. In practice sand resists moving around corners or projections and flowability varies from sand to sand, in general increasing with decreasing green strength, decreasing grain size, and with a change in moisture from the point of maximum strength.

(6) Mold hardness is determined by the degree to which a spring-loaded ball penetrates into the mold surface. A ball $\frac{1}{2}$ inch in diameter is connected to a dial gage which reads from 100 (no penetration) to 0 (0.10-inch penetration). A "hard-rammed" mold reads 90, a "soft" mold 50 to 60. Severe penetration by the liquid metal and washing of the sand occur when the hardness reading is below 50.

Miscellaneous additives. Although the principal properties of green sand are governed by the clay-water-silica relationship just discussed, various other materials are added for special purposes. Sea coal (a bituminous coal), pitch, or gilsonite (a mineral very rich in hydrocarbons) is added in the attempt to provide a reducing atmosphere which, as discussed later in the section dealing with mold-metal interface reactions, leads to better surface finish on the casting. At temperatures above its softening range, pitch also imparts some plasticity to the sand.

Cereal (corn flour) provides additional green strength through the formation of starch, and furthermore, has the advantage of charring and collapsing after being exposed to the heat of casting. Molasses and dextrin are used in a similar manner. These reactions can help prevent hot tearing and also facilitate shake out. Wood flour does not materially affect the bond in the sand but does improve the collapsibility.

8–10 Dry sand. Dry sand is a green sand modified by baking the mold at 400° to 600°F (204 to 316°C). The loosely bonded "mechanically held" water is thereby eliminated and only the tightly bonded water in the clay-water-silica bond remains, along with the water of crystallization in other mold materials. The heating also permits some flow of low-melting materials, such as pitch, and so provides better bonding.

Dry compression strength [11] (Fig. 8–11) is often affected by the amount of clay and also by the amount of tempering water that is used before drying. The behavior of the sodium montmorillonite seems distinctly different from that of calcium montmorillonite or kaolinite. This may simply be due to the fact that the dry compression tests yielding the data presented in Fig. 8–11 was limited to 100 psi. Above this value the

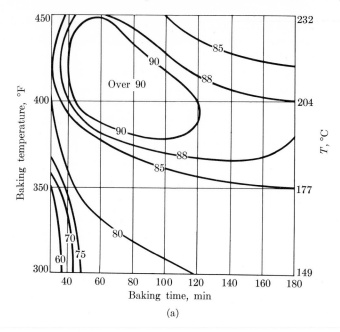

(a)

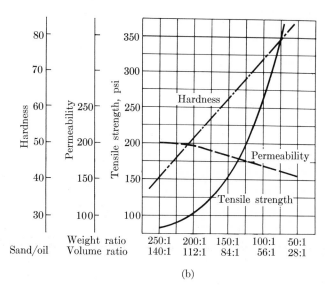

(b)

FIG. 8–12. (a) Effect of baking temperature and time upon dry compression strength. (b) Effects of processing variables on core-sand properties [15].

curves would certainly flatten out in all cases. The dry bond of the sodium montmorillonite is markedly higher than that of other clays. The residual bonding forces of the Na^+-to-water combination are evidently quite high.

8–11 Core sand. Core sand differs from dry sand in that only enough inorganic binder is employed to hold the core together before baking. The main source of strength is the organic bond developed after baking. The organic binders used include core oil, plastics, resins, and casein. In addition, cereal is used for both strength and collapsibility. The action of the principal binders involves polymerization and cross linking, as discussed in the section on organic bonds. For linseed oil, oxygen is the cross-linking agent, and it is therefore essential that sufficient air be circulated through the core oven.

Since a core is often surrounded by metal on several sides, it is imperative to provide adequate permeability to permit passage of gases to the atmosphere. For this reason, core sands are usually coarser than molding sands and are distributed over fewer sieves. Typical relations among processing variables are illustrated in Fig. 8–12.

CO_2 core process. The production of CO_2 cores differs from that of conventional cores, since no green strength is required. Before drawing, the core is bonded in place in the corebox by the action of CO_2 upon sodium silicate, as explained in the discussion of hydraulic bonds.

8–12 Shell sand. In shell molding, a dry mixture of sand and resin is dropped upon a hot metal pattern. The resin flows, bonding a "shell" of sand approximately $\frac{3}{16}$ inch thick adjacent to the pattern. The remainder of the mixture falls away, to be used for another mold. Upon continued heating the resin "sets." The commonest resin used is phenolformaldehyde, which is bonded before use by condensation polymerization (see organic bonding). The resin is then ground finely with a cross-linking agent, hexamethylenetetramine. In this state the product may be stored for long periods. Upon being heated, the resin flows around the sand grains, and with continued heating cross linking takes place.

A recent development is to precoat the sand by dissolving the resin in alcohol, mixing it with the sand, and then drying (the drying process drives off the alcohol). Upon application of heat during molding, the resin sets.

8–13 Investment molds. The principal feature of investment molding is that a liquid mixture flows around the pattern, giving excellent detail, and then sets with a "hydraulic" bond. In some cases the pattern is permanent and retractable; in others it is expendable and is removed by burning or melting.

The key to the process is the material used for bonding. For relatively low-melting metals, such as aluminum, plaster of paris is employed, while for higher temperatures silicate and phosphate bonds are used. The nature of the bonding has been discussed in detail under hydraulic bonds, and the molding process itself was considered in Chapter 7.

8–14 Permanent molds, die casting. Molds for these processes are usually made of metal, although semipermanent molds of graphite have been successful in some instances. The permanency of the metal mold is due principally to the ability of the metallic structure to resist heat shock. Although the coefficient of thermal expansion of metals is higher than that of most ceramics, uneven expansion does not lead to rupture since it can be compensated for by plastic flow. For this reason a metal mold may undergo distortion rather than crack. It is interesting to note the effect of uneven expansion on gray cast iron used for permanent molds which require pronounced shock resistance. Tiny ruptures actually occur near the graphite flakes in the microstructure, but gross distortion is avoided. A steel ingot mold, for example, exhibits marked distortions, whereas cast iron performs quite satisfactorily.

Thermal conductivity is an important property in metal molds. Copper, for example, can be used successfully as a mold for small steel castings. The rapid cooling rate results in the formation of a solid skin of steel before the steel has raised the temperature of the copper sufficiently to effect solution. Similarly, the high thermal conductivity of graphite is a significant factor in the successful use of this material.

REACTIONS BETWEEN LIQUID METAL AND MOLD
AND LADLE MATERIALS

8–15 Mold-metal interface reactions. Before leaving the problem of molding materials, we wish to review briefly the physical and chemical interactions between the mold and the liquid metal [12]. This is a relatively virgin field for research, because it requires a knowledge not only of the properties of the ceramic, organic, and metallic materials involved, but also of the various interactions that occur during the production process. High-temperature chemical reactions and such physical effects as changes in dimensions, which lead to spalling of the mold wall, must be considered. For centuries, foundrymen have been aware of this problem and have attempted to solve it by applying various "refractory" and "reducing" materials as washes to the mold wall.

As an example of a basic approach to this problem, let us consider liquid steel poured against a clay-bonded green-sand mold wall. We shall discuss first the changes taking place on each side of the interface and then the interaction of the materials. The changes on the green-sand side of the interface are illustrated by Fig. 8–6, which shows that quartz increases in volume with temperature, with a large expansion at 1063°F (590°C), and by Table 8–3, which records the thermal changes for clay, i.e. kaolin (these are similar to those exhibited by montmorillonite). The excess water is eliminated below 212°F (100°C), as shown by the weight loss of the clay.

On the metal side of the interface a layer of liquid steel is present for a short time which varies with the size of the section. This layer remains at relatively constant temperature for a long period of time because of the large heat block being delivered from the heat of solidification. (In the meantime, the ceramic side of the interface shows an increase in temperature.) The extent to which the existing sand surface is reproduced in detail depends upon the gas pressures and other reactions to be discussed later.

At the interface the clay is first broken down to an amorphous aluminum silicate mixture which in turn tends to form the equilibrium phases of silica (cristobalite) and mullite ($3Al_2O_3 \cdot 2SiO_2$). The silica is present partly as quartz and partly as cristobalite, the higher-temperature phase of silica.

The water vapor and the air oxidize the steel interface, thus producing iron oxide and other metallic oxides. The oxides flux the silica to form a relatively low-melting liquid iron silicate, as described by the SiO_2-FeO-Fe phase diagram, and this liquid silicate wets the silica grains rapidly. At the same time, the liquid metal, which does not wet the silica but does adhere to the silicate, penetrates the sand and is further oxidized. This penetration continues below the original melting point of the steel, since dissolved oxygen lowers the melting point. In this context it should be noted that even a small depression of the melting point, amounting to a few degrees only, will cause some liquid metal to be present at the interface for a relatively long period of time, due to the fact that the temperature is kept fairly constant by the heat of solidification of the unoxidized metal farther back in the mold. Upon further cooling the liquid silicate crystallizes to fayalite, Fe_2SiO_4, and is observed in this form in the final casting.

From this brief description it is evident that both the quality of the casting surface and the adherence to dimensional tolerances are vitally affected by the mold-metal interface reactions.

The effect of the mold atmosphere on the interface reactions has re-

TABLE 8–3

THERMAL CHANGES IN KAOLIN [1]

Temperature		Shrinkage rate	Porosity	Heat effects	Weight loss	Petrographic results	X-ray difffraction results
°C	°F						
20–100	68–212	zero	high	none	slight	kaolinite crystals	kaolinite crystals
100–400	212–752	zero	high	none	slight	kaolinite crystals	kaolinite crystals
400–500	752–932	slight	high	large absorption	very large	breakup of crystals	breakup of crystals
500–900	932–1652	medium	high	none	slight	no visible crystals	amorphous meta-kaolin
900–1000	1652–1832	high	medium	large evolution	none	no visible crystals	γ-alumina and mullite appear
1000–1150	1832–2102	zero	medium	none	none	growth of mullite	growth of mullite
1150–1200	2102–2192	high	rapid decrease	small evolution	none	formation of cristobalite	mullite and cristobalite

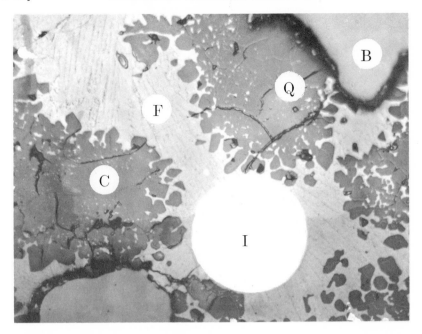

Fig. 8–13. Microstructure developed in samples of pure quartz sand heated in contact with pure iron particles for 60 minutes at 2850°F (1566°C), in an atmosphere of 10% CO_2 and 90% CO. Q = quartz, I = iron, B = bakelite mount, C = cristobalite, F = fayalite. (250X.)

cently been investigated [12]. Samples of pure quartz sand were heated in contact with fine iron particles at temperatures above and below the melting point of iron and in atmospheres in which the CO/CO_2 ratio was adjusted to yield varying degrees of oxidation. In an atmosphere containing 10% CO_2 and 90% CO, the liquid iron was converted to oxide at 2850°F (1566°C). The oxide fluxed the silica, forming a glassy melt from which fayalite (Fe_2SiO_4) crystallized during cooling (Fig. 8–13). At 2.7% CO_2 and 97.3% CO, the iron was not attacked (Fig. 8–14). At lower temperatures, such as 2233°F (1227°C), the iron was not attacked by a gas mixture containing 10% CO_2, but was severely attacked by a gas consisting of 50% CO_2 (Fig. 8–15). A summary of the data obtained by means of additional calculations [16] is presented in Fig. 8–16. The important region of this diagram is the curve in the lower right-hand corner indicating the field (metal and silica) in which no attack occurs at the temperatures and CO_2/CO ratios indicated. If the temperature or oxidizing potential of the gas is increased to reach a point in the field (melt and silica), then the iron will be oxidized to a glassy melt.

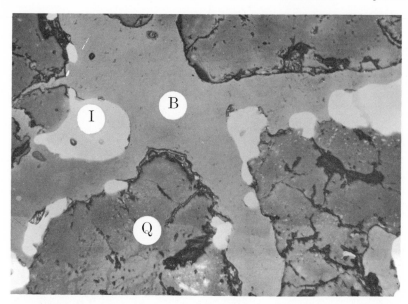

FIG. 8–14. Microstructure in reducing atmosphere (2.7% CO_2, 97.3% CO) Q = quartz, I = iron, B = bakelite. (No silicate melt or reaction products.) Specimen heated for 60 minutes at temperature of 2777°F (1525°C). (250X.)

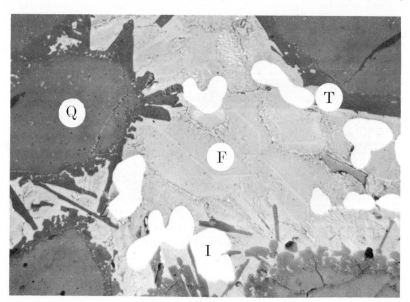

FIG 8–15. Microstructure developed in oxidizing atmosphere (50% CO_2, 50% CO) at lower temperatures (2233°F or 1225°C for 60 minutes). Q = quartz, I = iron, T = tridymite, F = fayalite. (250X.)

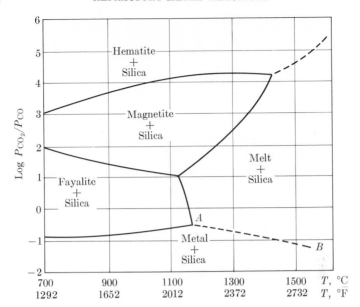

FIG. 8–16. Experimental locations for specimens superimposed on Darken's [8] equilibrium diagram. (Fe-Si-O system.)

Another significant feature shown by the photomicrographs of Fig. 8–15 bears upon the problem of metal penetration into the mold—an effect which produces a poor casting surface. Note that the liquid iron in Figs. 8–13 and 8–14 does not wet the silica and hence does not tend to penetrate it except at high ferrostatic pressures. On the other hand, the melt formed by the reaction of iron oxide and silica wets the silica and opens up channels through the mold wall for the metal to flow. Therefore, both penetration and reaction at the mold-metal interface can be reduced by minimizing the oxidizing potential. This is probably an important reason why good surfaces are produced by using either dry sand, cores, or shell molding. The water vapor of green sand provides a higher oxidizing potential, and in heavy castings there is an opportunity for extensive interface reaction.

8–16 Refractory-metal reactions. Closely allied to the mold-metal reactions are the reactions between the liquid metal and the refractories of the furnace and ladle. Since cast steel is melted and handled at higher temperatures than any of the other common cast alloys, an example of its reactions with refractories will illustrate the problem most clearly. For many years, serious defects called "ceroxides" or nonmetallic macroinclusions have been encountered in certain steel castings, particularly in

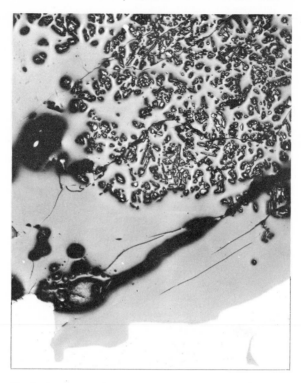

FIG. 8–17. Typical cross section through metal (bottom) and macroinclusions (top). Corundum crystals in a glassy matrix (top) are followed by glass and then cast steel (white spots).

those with large flat cope surfaces molded in green sand. Petrographic examination of these large inclusions disclosed that they contained the mineral corundum, Al_2O_3, in a glassy matrix (Fig. 8–17). It was determined [17] that this phase originated from the reaction between the aluminum (added for deoxidation of the steel) in the liquid steel and the fireclay of the ladle refractories. To minimize the attack, either high alumina or magnesia refractories should be used in the ladles.

REFERENCES

1. F. H. NORTON, *Elements of Ceramics*. Reading, Mass.: Addison-Wesley, 1952.

2. G. J. VINGAS and A. H. LEWIS, "Anionic or Cationic Agents—A Solution to Sand Problems," *Trans. A.F.S.* **64**, 453 (1956).

3. B. H. BOOTH, "Sand Properties vs. pH," *Trans. A.F.S.* **57**, 210 (1949).

4. W. DAVIES, *Foundry Sand Control*. The United Steel Companies, United Kingdom (1950).

5. V. E. ZANG and G. J. GROTT, "Casting Quality as Related to the pH Value of Molding Sands," *Trans. A.F.S.* **62**, 393 (1954).

6. V. E. ZANG and G. J. GROTT, "Use of Engineering Methods in Practical Sand Problems; Strength Density Relationships," *Trans. A.F.S.* **63**, 383 (1955).

7. A. H. ZRIMSEK and R. W. HEINE, "Clay, Fines and Water Relationships for Green Strength in Molding Sands," *Trans. A.F.S.* **63**, 575 (1955).

8. H. W. DIETERT, "Processing Molding Sand," *Trans. A.F.S.* **62**, 1 (1954).

9. MANUEL F. DRUMM, "Resins and Sands for Shell Molding," *Trans. A.F.S.* **61**, 367 (1953).

10. H. F. TAYLOR and L. M. DIRAN, "The Nature of Bonding in Clays and Sand Clay Mixtures," *Trans. A.F.S.* **60**, 356 (1952).

11. RALPH E. GRIM and F. L. CUTHBERT, "The Bonding Action of Clays," University of Illinois Bulletin, Series No. 357, Vol. 42, No. 50, No. 362, Vol. 43, No. 36 (1945).

12. G. A. COLLIGAN, L. H. VAN VLACK, and R. A. FLINN, "The Effect of Temperature and Atmosphere in the Iron-Silica Interface Reaction," *Trans. A.F.S.* **66**, 452 (1958).

13. C. PALACHE, H. BERMAN, and C. FRONDEL, *Dana's System of Mineralogy*. Vol. 2. New York: Wiley, 1956, p. 484.

14. C. PALACHE, H. BERMAN, and C. FRONDEL, *Dana's System of Mineralogy*. Vol. 1. New York: Wiley, 1944, pp. 4–5.

15. H. W. DIETERT, *Foundry Core Practice*, A.F.S., 1950.

16. L. S. DARKEN, "Melting Points of Iron Oxide in Silica; Phase Equilibria in the System Fe-Si-O as a Function of Gas Composition and Temperature," *J. Am. Chem. Soc.* **70**, 2046 (1948).

17. L. H. VAN VLACK, R. A. FLINN, and G. A. COLLIGAN, "Non-Metallic Macroinclusion Causes in Steel Castings," *Trans. A.F.S.* **68**, 132 (1960).

18. L. H. VAN VLACK, *Elements of Materials Science*. Reading, Mass.: Addison-Wesley, 1959.

GENERAL REFERENCES

Transactions of the American Ceramic Society.

PROBLEMS

1. A silica brick is composed of a mixture of 50% cristobalite, 25% tridymite, and 5% glass (all are SiO_2). How much thermal expansion (% volume) occurs between 68°F (20°C) (room temperature) and 660°F (350°C)?

2. The compound $AlPO_4$ can exhibit the same crystal structures as SiO_2. Explain this phenomenon, using the data of Table 8–1.

3. Why is SiO_2 glass deformable at high temperature although MgO is not?

4. Why are plaster of paris investments used for aluminum and not for steel castings?

CHAPTER 9

MOLD DESIGN AND PROCESSING METHOD.
OPTIMIZATION OF CASTING DESIGN.

9–1 General. Up to this point we have given separate consideration to such important features of casting as the flow and solidification of liquid metal, the stresses developed in the solid state, the salient factors in mold design, and the molding materials and processes available. To consolidate our position at this time, it seems worthwhile to indicate how all these separate pieces of information may be marshalled to guide the engineer confronted by the following two cases of casting production:

(1) *Inflexible casting design.* When the casting design is fixed, the choice of mold design and process must be adapted to the specifications.

(2) *Flexible casting design.* The casting design can be optimized to provide maximum strength and ease of production.

In practice, the situation often lies between these extremes, and minor alterations in processing are usually allowed. In discussing these two cases, we shall also present a summary of casting defects and the methods used to prevent them, to provide a background for some of the points covered in Section 9–9, which deals with the problem of optimizing casting designs.

9–2 Choice of molding materials and methods for a given design. The choice of molding materials and processes is governed largely by the following factors (see Table 9–1): (1) casting geometry (maximum and minimum dimensions); (2) casting tolerance (surface finish, closeness of dimensions); (3) metal composition; (4) number of castings required.

(1) *Casting geometry.* From Tables 9–1 and 9–2, it is evident that some of the methods are limited by the weight of the casting. To the enterprising engineer this should not be an insurmountable barrier. If the advantages of a given process are very important for a certain casting, a way can usually be found to adapt the process to the particular application. As a general rule, however, we would not expect to see investment methods used to pour a 10-ton machine base. Although such a part has to be produced to certain accurate dimensions, it is far more economical to obtain these by machining.

Other limitations listed in Tables 9–1 and 9–2 may also need further explanation. For instance, the problem of minimum section is very important in many aircraft parts, and in these cases it is possible to produce lighter sections than those listed, by using higher pouring temperatures, air pressure, or centrifugal casting methods. An alternative in

185

TABLE 9–1

COMPARISON OF MOLDING METHODS, MATERIALS, AND ALLOY COMPOSITIONS

Group	Process	Pattern material	Mold material	Alloys cast	Size range, lb	Average size, lb	As-cast finish, rms	Labor cost/lb	Equipment cost/lb	Mold material cost
I Pattern and mold expendable	Lost wax or investment	Wax and polystyrene	Silica base ceramic* bond	No limitations (however, low carbon steel is difficult)	0.01-5	0.2	50	high	high	high
	Frozen mercury	Frozen mercury			0.1/100	1	50	high	high	high
II Permanent pattern, expendable mold	Green sand Dry sand Core sand	Wood or metal	Green sand	No limitations	Not limited	1/50	100	low	low	low
	Shell	Metal	Dry sand and resin	No limitations	0.2/80	1/80	100	low	medium high	medium
	Plaster	Plastic, metal	Plaster	Al, Mg, some Cu base. Not ferrous	1/50	1/10	70	medium	medium high	medium
III Permanent mold	Permanent mold	None or master pattern	Metal or graphite	No limitations (however, steel is difficult)	1/500	1/10	120/200	low	high	low
	Die casting	None or master pattern	Metal	Al, Mg, Zn base. Some Cu base	0.1/75 lb Al 0.2/200 lb Zn	0.1/1	70	low	high	low

* From ethyl-silicate or magnesium-phosphate cementing reaction.

some cases is to cast heavy sections in such a way that an inexpensive machining operation can be used to remove surplus material.

Not all the limitations of the processes are listed in the tables. In shell molding, for example, the largest pattern plate which can be accommodated on a machine at present is 40 × 36 inches. Therefore, it would be difficult to pour a casting exceeding these dimensions even though its weight might be less than 200 pounds.

(2) *Casting tolerances.* Minimum tolerances should be specified only when absolutely necessary, since holding all portions of a casting to these dimensions adds to pattern and production costs. If a part is to be machined in a given region, it is usually better to provide sufficient stock to produce a proper machining chip than to reduce chip thickness. In a heavy-section steel casting the possibility of a "burnt-on" sand condition is great, and it is better for the cutting tool to pass well beneath this layer than scrape through it. The improved tool life and better surface finish far outweigh the cost of the greater amount of material removed.

The greatest saving deriving from such precision casting methods as investment, shell, die, and permanent molds is realized when certain machining operations can be eliminated rather than merely reduced. Grinding, however, is an exception to this rule, particularly when hard or abrasive alloys are being processed, because grinding time is almost directly proportional to the amount of material removed.

(3) *Metal composition.* A survey of Tables 9–1 and 9–2 shows that certain methods, such as die and plaster casting, are not applicable to metals with high melting points. The iron-base alloys are a case in point. Plaster casting is not suitable for these materials because the mold-metal reaction between plaster and steel is quite severe and, as was discussed in Chapter 8, mold-metal interface reactions can be as important as temperature.

(4) *Number of castings required.* Many methods require expensive metal patterns or dies. If only a few castings are required, the use of these methods leads to a very high cost per casting.

9–3 Balancing costs. Although the question of cost comparison among different methods often arises, it cannot be answered in general terms. Each case requires study by personnel who understand the methods, and even after casting costs have been compared, the total manufacturing cost, including finishing, as well as the service life of the finished product, must be taken into account. Other aspects of the cost survey are rate of production, lead time, reliability of delivery, and allowable rate of depreciation of equipment. For example, if it is desired to produce a small part to accurate dimensions by investment casting, the quantity should justify the expense of preparing accurate dies for wax or plastic patterns.

TABLE 9–2

MINIMUM SECTION THICKNESSES AND DIMENSIONAL TOLERANCES FOR DIFFERENT CASTING PROCESSES AND CAST METALS [2]

Material*	Green or dry sand				Shell mold				Permanent mold			
	Minimum section for length		Dimensional† tolerance for casting ± in.		Minimum section for length		Dimensional† tolerance for casting ± in.		Minimum section for length		Dimensional† tolerance for casting ± in.	
	1 in.	6 in.	1 in.	6 in.	1 in.	6 in.	1 in.	6 in.	1 in.	6 in.	1 in.	6 in.
Aluminum alloys	0.125	0.140	0.015	0.015	0.062	0.120	0.005	0.013	0.100	0.160	0.007	0.010
Copper alloys	0.090	0.140	0.015	0.015	0.090	0.120	0.005	0.013	0.100	0.200	0.015	0.021
Gray irons	0.125	0.200	0.030	0.050	0.125	0.150	0.010	0.030	0.250	0.400	0.030	0.030
Steel	0.250	0.250	0.060	0.100	0.180	0.250	0.010	0.030	0.250	0.250	0.010‡	0.020‡

Table 9–2 (continued)

Material*	Die cast				Investment or plaster				Centrifugal			
	Minimum section for length		Dimensional† tolerance for casting		Minimum section for length		Dimensional† tolerance for casting		Minimum section for length		Dimensional† tolerance for casting	
			± in.				± in.				± in.	
	1 in.	6 in.	1 in.	6 in.	1 in.	6 in.	1 in.	6 in.	1 in.	6 in.	1 in.	6 in.
Aluminum alloys	0.050	0.080	0.005	0.080	0.060	0.060	0.005	0.007	0.070	0.150	0.010	0.012
Copper alloys	0.030	0.060	0.005	0.020	0.040	0.060	0.005	0.010				
Gray irons												
Steel					0.060	0.060	0.005	0.015				

* Magnesium base and zinc-base alloys are cast to tolerances equal to those applying to the aluminum alloys.
† The following additional tolerances should be applied if the dimension chosen crosses the parting: sand ± 0.015, shell mold ± 0.010, permanent mold ± 0.015, plaster ± 0.010, centrifugal ± 0.010 for a one-inch casting. Similar tolerances should be added if the dimension is between a core and a mold surface.
‡ Pressure poured in graphite mold.

9–4 Fluidity and gating. Many strong claims have been made concerning the advantages of special molds or of heated molds in promoting fluidity. Certain mold coatings such as amorphous carbon and hexachlorethane have been found useful in light metals [2]. For example, extremely intricate detail can be obtained in silver or gold alloys cast in investment molds when the mold temperature is within several hundred degrees of the freezing range. Similarly, in the die casting of such low-melting alloys as the zinc-base group, the mold temperature can have a dominant effect. In general, the fluidity (castability) of an alloy will vary as a function of the superheat above the liquidus, as described in Chapter 2, but the slope of the curve may vary with mold temperature and material.

The novice is often amazed to discover how successfully misruns in permanent molds of relatively high thermal conductivity can be prevented by controlling certain variables that affect metal delivery. For example, increased pouring rates, acceleration by pressure, streamlined gating systems, and special mold washes make it possible to pour at temperatures well below those commonly used for equivalent sand castings.

Gating principles are only slightly affected by the choice of mold material and method. As mentioned earlier, the lost-wax method permits the use of more elaborate gating systems. There is scattered evidence that the smoother gate passages obtained by the "precision" methods may improve the rate of flow somewhat, but not enough to influence seriously the gating calculations already discussed.

9–5 Risering. Risering calculations are affected greatly by the choice of mold material. The effect of substituting metal or graphite for a sand surface has been discussed previously. One of the great advantages of the permanent-mold and die-casting methods is the rapid solidification rate which converts a wide freezing band to a narrow band for a given alloy. The quantitative effects of chilling have already been discussed in Chapter 3. At this point it remains only to consider the possibility of using chills in the different molding methods.

In the investment processes, Type I of Table 9–1, insulating materials or chills are rarely used. In the investment-shell technique, it is possible to back up the shell with either metal or shot or insulating material, to develop favorable thermal gradients. In the standard investment technique one might expect the thermal gradients to be smaller than in a sand mold of the same design because of the high mold temperature at pouring (1600 to 1800°F, 871 to 982°C).

In Type II, the sand-molding processes, chills find the widest application. Inasmuch as the molds are never heated above 500°F (260°C) in any of the baking processes, no oxidation problems arise. Also, in copper-base alloys insulated risers have been used with success.

In Type III, the permanent-mold processes, the entire casting is chilled unless insulation is provided. For the manufacture of steel car wheels in graphite molds, a very effective method of riser insulation uses the heated mold to bake a thin annulus of shell sand around the riser cavity. In permanent-mold and die casting, insulating washes have proved effective. All these methods are based on the great difference between the cooling rate of sand and that of metal, which permits an appreciable reduction in the cooling rate by means of a relatively thin wash.

9–6 Internal stresses and defects. The basic reasons for hot tearing and residual stresses were described in Chapter 6. We shall now consider the various degrees to which the different molding materials and processes are susceptible to these defects.

Investment castings are least susceptible to both hot tears and residual stresses because the molds are usually poured at high temperatures (above 1600°F, 871°C) so that the mold expansion is markedly reduced while the casting is contracting during solidification. Residual stresses are also minimized because the entire assembly of investment and metal reaches equilibrium temperature shortly after pouring, i.e., at a time when the metal is usually still in the plastic range. No large thermal gradients are present when the casting passes into the elastic range, and therefore only minimal residual stresses are encountered.

Permanent-mold castings are very susceptible to hot tears because of the unyielding nature of the mold. The metal mold also responds to heating faster than sand and so undergoes severe expansion while the casting is cooling. These factors require careful consideration by the mold designer in order to avoid a clamping action in certain sections of the casting (for example, in a U-section). For these reasons, gates are made very thin to afford easy fracture, and extensive use is made of sand cores.

In die casting, it is customary to adopt the highest die temperatures that will give acceptable die life and production rate.

9–7 Cause and cure of common casting defects. Although our emphasis throughout has been on methods leading to satisfactory castings, it will nonetheless be helpful to summarize typical casting defects, and to discuss causes and remedies. Two causes of casting rejects will not be dealt with: (a) scrap caused by deviation from metallurgical specifications, such as off-composition and high chill, and (b) scrap caused by gross negligence, such as the use of wrong cores. In both instances, the remedies are obvious.

The remaining defects can be grouped into two categories: those primarily due to the mold and those caused principally by liquid metal delivered at the wrong temperature or rate. We shall briefly describe the various types of defects that occur in each group.

The following defects are due to the mold.

(a) *Cuts and washes* result from erosion caused by the mechanical action of the liquid metal, combined with thermal spalling. In general, these defects are found when narrow gates are handling large volumes of metal over a relatively long pouring time. Refractory inserts or a change in mold design is indicated.

(b) *Scabs, buckles, and rat tails* are terms used to describe the breakdown of the mold wall under thermal shock. In the case of a *scab*, a portion of the mold surface has broken out and the recess has filled with metal. Therefore, the casting must be carefully inspected on the cope surface for sand inclusions. Such a defect calls for a sand with higher thermal-shock resistance or a change in mold design. A *buckle* is really an incipient scab; the mold surface has spalled and the pieces have not yet fallen out. A *rat tail* is a minor buckle.

(c) *Fusion and penetration* are discussed together because they are often confused. *Fusion* of the mold sand results in a glassy layer of burnt-on sand. There is no metal between the sand grains, merely a bond of the sand to the metal. This defect is common in metals poured at high temperatures, such as steel. Even when pure sands and clays are employed, the iron oxide generated at the surface of the steel at the high interface temperature can act as a flux for bonding. *Penetration* is the passage of liquid metal through the sand grains, and the result is an intimate mixture of metal and sand. Here, the cure is a finer sand, better ramming, lower pouring temperatures, or reduced mold-metal reaction (the last alternative calls for a less reactive sand or metal).

(d) *Crushes, drops, ramoff, stickers, swells, shifts, runouts, core raise, sand inclusions* are types of defects which are usually due to poor molding practice and are easily eliminated by adequate supervision.

Crushes result from mold surfaces that have been damaged either by misalignment of flasks or improper core setting. *Ramoff* is a protuberance on the casting caused by movement of the pattern during ramming, thus producing an oversize cavity at one region. *Stickers* occur when a mold has been damaged by sand that has stuck to the pattern. *Swells* are caused by abnormal expansion of the mold during pouring. Usually, soft ramming results in a mold surface incapable of withstanding the hydrostatic pressure. In gray iron and ductile iron, the expansion taking place during solidification also causes swelling. Rigid molds eliminate the defect and produce sounder castings. *Shifts* result from a misalignment of cope and drag, and may be due to faulty pins or poor mold assembly. *Runouts* are developed by improper mold clamping or by an unsatisfactory sand seal between cope and drag. *Core raise* is a term used to describe the tendency of a core submerged in liquid metal to float unless properly anchored either by the print, chaplets, or cement. *Sand inclusions,* which may be due to any one of the failures discussed above, cause sand to fall into the liquid metal. Also, if molding practices are poor, loose sand may be found in the mold cavity or gate passages. This sand can usually be distinguished from slag by its texture which is grainy rather than glassy.

(e) *Hot tears* have been discussed at length in Chapter 6; the effect of mold restraints should be reviewed in this regard.

(f) *Blows, scars and plates* are defects arising from the action of gas. A *blow,* caused by the rapid evolution of gas from a wet mold or chill, produces gas holes in the metal. By contrast, mold gas may build up in cope pockets in amounts sufficient to prevent metal flow. Such an unfilled pocket is a *scar.* Sometimes platelike channels of metal, called *plates,* will ooze into the pocket.

(g) *Shrinks* or shrinkage cavities have already been discussed at length. They are due to an inadequate risering system incapable of feeding the liquid-to-solid shrinkage. High pouring temperatures can aggravate shrinkage.

The following defects are due to improper metal or mode of delivery.

(a) *Gas porosity* is a condition characterized by a group of voids due to gas which is dissolved in the metal and liberated during cooling (compare with *blows* above). Note that this gas is not mechanically entrapped (see detailed discussion in Chapter 10).

(b) *Misruns, cold shuts, shot iron.* Excessively low pouring temperature and rate of metal delivery produce *misruns* (improperly filled corners and mold cavities) and *cold shuts* or *seams* (defects at the junction of two streams of metal). *Shot iron* is the term used for pellets of iron found in a fractured casting section. These pellets form when a vertical stream breaks up into particles which freeze separately and are entrapped in the large mass of metal without fusing. Improved gating and higher pouring temperatures are remedies.

(c) *Slag.* This term encompasses several defects caused by gross ceramic inclusions which may belong to one or more of the following categories: (i) furnace slag, (ii) oxides and silicates formed by deoxidation or other ladle treatment, (iii) ladle refractories which are fluxed or melted by the metal, and (iv) sand from the gating system entrapped in the metal stream. In correcting defects of this type it is quite important to determine the source of the slag. Defects due to slag are also often accompanied by gas holes caused either by the reaction of the slag with metal (for example, iron oxide with carbon to give carbon monoxide) or by the nucleating effect of the inclusions on bubble formation.

9–8 Optimizing casting design for maximum strength and ease of processing. So far we have considered the casting design as invariable and developed the mold design to fit the casting. Unfortunately, this approach is frequently necessary since the engineer is quite often confronted with situations in which the pattern has already been made and no modifications are allowed. However, in the ideal case, the component design is the result of a cooperative effort between the engineering staffs of the user and the foundry, where all the features of the casting process are taken into account. This procedure pays two highly desirable dividends:

(1) Improved *component design* (due to the streamlining possible in the casting process which is used to minimize stress concentrations).

(2) Improved *processing* of the casting (due to the elimination of costly cores, bosses and risers).

Although there are thousands of examples confirming the point just made, we will discuss only a few principles to illustrate the importance of keeping the features of the casting process in mind during the design stage. We will see that in many cases the improved design also leads to simplified processing.

In our discussion we shall follow a path from simpler concepts to more complex matters, i.e., we shall observe the following order: (1) minimum section thicknesses for different alloys; (2) variations in section thickness; (3) junctions between sections; (4) section shapes; (5) the casting as a whole (parting, cores, chills).

(1) *Minimum section thicknesses.* For many components in which weight is important, or where a surface giving maximum heat transfer is required (for example, in fins for air-cooled cylinder heads), the question of minimum section has to be considered. Table 9–2 lists the section thicknesses which are normally poured in the various alloys without extra cost. The data are intended as a general guide. It is possible to produce sections of one-half these thicknesses by means of special techniques such as higher pouring temperatures and application of pressure during pouring. In general, it is evident that the lower-melting alloys are the metals to be used for thin sections. In sand castings, for example, $\frac{1}{8}$-inch sections are common for aluminum, whereas in steel the minimum is usually $\frac{1}{4}$ inch. However, in investment castings, the highly preheated molds and other special techniques permit in steel sections as thin as 0.06 inch.

(2) *Variations in section thickness.* A great deal has been written about the desirability of designing castings with uniform sections. If two heavy sections are joined by a light section, it becomes necessary to riser each heavy section separately. This requirement is eliminated if the design calls for uniform sections. It should be pointed out, however, that

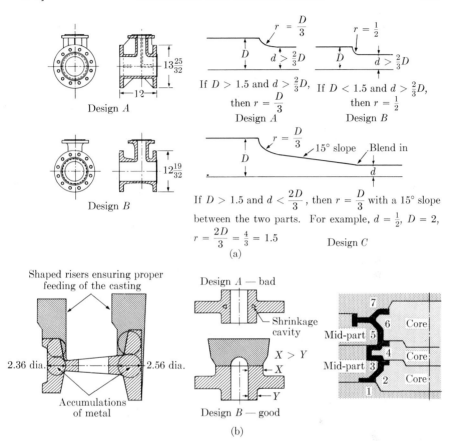

(a)

(b)

Fig. 9–1. (a) Designs [defective (A) and correct (B)] for a T-piece for steam-pressure service. (b) Schematic representation of pouring technique for a wheel casting, demonstrating principle of nonsymmetry and the process of molding a complicated shape.

sections tapered and enlarging toward the source of feed metal are often desirable. The T-piece for steam-pressure service, Fig. 9–1(a), is a good example. Note that the heavier masses are at the ends of the part where they can be risered. The hub (Fig. 9–1b) is a similar case, while the wheel is an example of an expensive design. In joining a heavy to a light section, the recommendations sketched in Fig. 9–1 should be followed. In many current casting designs, these principles have not been applied. Instead the design may have been translated directly from an assembly of rectangular shapes or drawn without thought of the importance of reducing stress concentrations.

(3) *Junctions between sections.* In designing the various types of junctions, the engineer should remember that greater casting strength

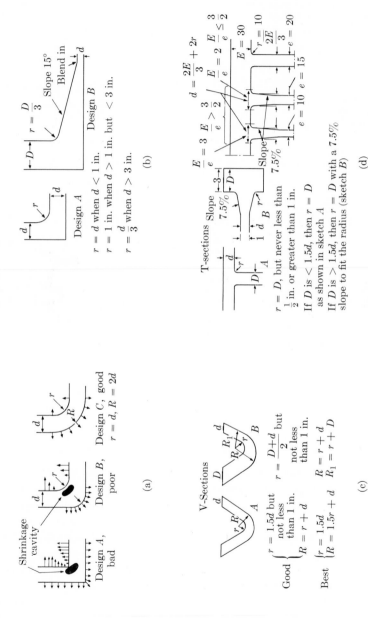

Fig. 9-2. (a) Evolution of the L-design, eliminating shrinkage. (b) Dimensional relationships of L-design with exterior corners. (c) Recommended V-design relationships. (d) Recommended T-design (left) and T-junction joining unequal walls. (*cont.*)

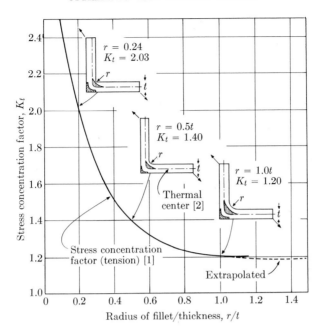

FIG. 9–2. (*cont.*) (e) Relationships between stress concentration of an L-section in tension and thermal gradients near the end of solidification [1 and 3].

can be obtained if certain traditional, easy-to-draw shapes are avoided. The problem presented by the junction of structural members coming from different directions becomes more difficult as the number of members increases. The joint of two members can be shaped like the letters L or V, depending upon the angle of the junction, that of three members will be Y- or T-shaped, and the joint of four members will look like an X. However, if possible, no more than three members should ever be joined at one location.

An L-junction made with a sharp corner (design A of Fig. 9–2a) is subject to both severe stress concentration and shrinkage. These undesirable effects can be avoided by providing a uniform section or at least the proper radius for the inner corner. (For joining unequal sections, the gradual taper is recommended.) The stress concentration factor K_t (Fig. 9–2c) is the familiar multiplication factor to be applied to the average stress for the section. The shaded area at the centerline of each section represents metal which is still liquid 50 seconds after pouring. The section thickness is 4 inches (0.3% C steel). Note that for the sharp-cornered design, this zone of potential shrinkage is close to the surface. Thus, if a sharp internal corner is present, there are two dangers: stress concentration and potential shrinkage voids near the surface.

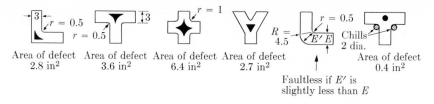

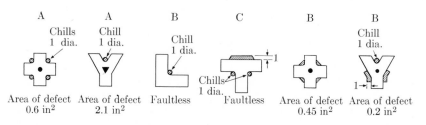

FIG. 9–3. Defects in the principal types of web intersections, demonstrating superiority of L- and T-junctions. A: defective junction. B and C: improved junction [1].

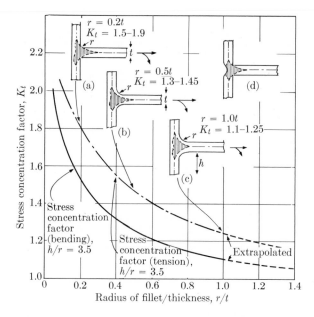

FIG. 9–4. Relationship between stress concentration and thermal gradients of T-junction [3].

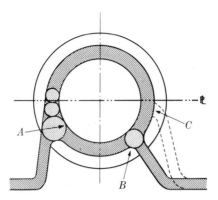

FIG. 9–5. Cross section of a motor-frame casting, showing defective and improved junctions and conversion of Y- to T-section [1].

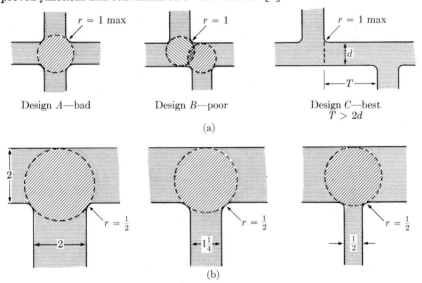

FIG. 9–6. (a) Design of X-sections. (b) Design of T-sections [1].

These problems can be avoided by following the design recommendations of Figs. 9–2(a) and 9–3.

The V-intersection intensifies the problem presented by the L-junction because the heat transfer is restricted even more at the inner point of the V. The remedy is to provide a generous radius as illustrated by design B in Fig. 9–2(b).

T- and Y-shaped sections are more difficult to handle than L- and V-shapes because the increase in the corner radius is accompanied by an increase in the size of the hot spot in the center (Fig. 9–4). The Y is

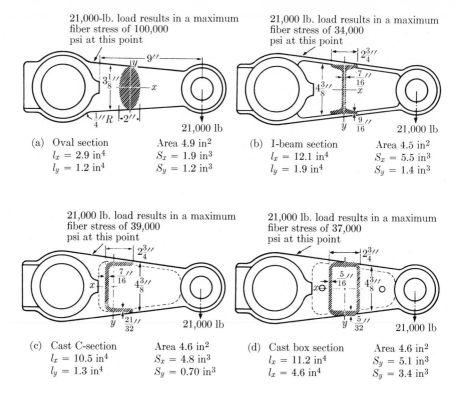

21,000-lb. load results in a maximum fiber stress of 100,000 psi at this point

21,000 lb. load results in a maximum fiber stress of 34,000 psi at this point

(a) Oval section   Area 4.9 in^2
     $l_x = 2.9$ in^4 $S_x = 1.9$ in^3
     $l_y = 1.2$ in^4 $S_y = 1.2$ in^3

(b) I-beam section   Area 4.5 in^2
     $l_x = 12.1$ in^4 $S_x = 5.5$ in^3
     $l_y = 1.9$ in^4 $S_y = 1.4$ in^3

21,000 lb. load results in a maximum fiber stress of 39,000 psi at this point

21,000 lb. load results in a maximum fiber stress of 37,000 psi at this point

(c) Cast C-section   Area 4.6 in^2
     $l_x = 10.5$ in^4 $S_x = 4.8$ in^3
     $l_y = 1.3$ in^4 $S_y = 0.70$ in^3

(d) Cast box section   Area 4.6 in^2
     $l_x = 11.2$ in^4 $S_y = 5.1$ in^3
     $l_x = 4.6$ in^4 $S_y = 3.4$ in^3

21,000 lb. load results in a maximum fiber stress of 44,000 psi at this point

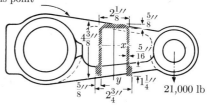

(e) Cast U-section   Area 4.6 in^2
     $l_x = 9.1$ in^4 $S_x = 4.2$ in^3
     $l_y = 6.2$ in^4 $S_y = 5.0$ in^3

FIG. 9–7. Effect of design on the stress in a lever and the lever's load-carrying ability [4].

inferior to the T; the conversion of a Y- to a T-section is shown in Fig. 9–5. The chilling of a T to produce soundness is illustrated in Fig. 9–3.

The X-section and other complex sections should be redesigned to call for a maximum of three members at the intersection (Fig. 9–6).

(4) *Section shapes.* In his choice of junctions, the designer is often governed by his familiarity with shapes such as the I-beam which he has encountered in the design of rolled products. From our preceding discussion, it is evident that this shape consists of two undesirable T-intersections. However, various other shapes are available. For example C-, U-, and box-shaped sections can be used, as illustrated in Fig. 9–7. The C- and U-sections may be cast without the cores required for the box section. Detailed stress calculations are given in reference 4.

(5) *The casting as a whole.* Casting designs seem almost endless in variety, and the beginner is often confused by their complexity. Although it takes time to become familiar with the wide spectrum of processing techniques, several suggestions may be offered. First, the most complex casting may be dissected into simple bar, plate, and cube sections and treated accordingly. It is often possible to obtain a better understanding of the flow pattern of the metal by proceeding with a little ingenuity, such as pouring only one-half or one-third the total volume of metal required into the mold and then opening the mold carefully. To investigate the relative cooling rates of different sections, thermocouples may be embedded or, even more simply, a casting may be shaken out of the mold while hot. Either visual observation for ferrous alloys, or temperature sensitive crayons (Tempilstik) may be employed to determine the approximate gradients. Finally, to ascertain the residual stresses or stresses under service loads, the brittle lacquer and strain-gage techniques may be employed (see general references).

Specific References

1. *A.S.M. Handbook*, vol. 1. A.S.M., Cleveland, Ohio, 1961, pp. 140, 893, 978.

2. M. C. Flemings, H. F. Conrad, and H. F. Taylor, "A Fluidity Test for Aluminum Alloys; Tripling Fluidity with a New Mold Coating," *Trans. A.F.S.* **67**, 496 (1959).

3. J. B. Caine, "Interrelation Between Stress Concentration and Castability," *Trans. A.F.S.* **67**, 121 (1959).

4. J. B. Caine, "Design Properties of Four Streamlined Cast Sections," *Foundry* 166–176, (April, 1959) 92–95, (May, 1959).

GENERAL REFERENCES

F. A. BRANDT, H. F. BISHOP, and W. S. PELLINI, "Solidification at Corner and Core Positions," *Trans. A.F.S.* **61**, 451 (1953).

Steel Castings Handbook. Steel Founders Society of America, Cleveland, Ohio, 1961.

Gray Iron Castings Handbook. Cleveland: Gray Iron Founders Society, 1958.

"Stress Analysis." *Transactions of the Society for Experimental Stress Analysis.* 1948 to present.

W. R. OSGOOD, *Residual Stresses in Metals and Metal Construction.* New York: Reinhold, 1954.

PROBLEMS

1. Select some component which is now made by another process and redesign for economical casting. For example, redesign a welded machine base shown in an advertisement.

2. Show locations in which defects might have been encountered in the original design illustrated in the advertisement, and indicate how these have been eliminated in your new design.

PART II

The Chemistry of Liquid Metal; Control of composition, melting, and refining

CHAPTER 10

GASES IN METALS

10–1 General. One of the most significant and progressive trends in the metal-casting industry is exemplified by the changes that have taken place in the melting operation. The old philosophy was that the foundry existed to melt and cast—it was up to the purchasing department to specify the proper metal composition which, moreover, was confined to a very few elements. In those days, the most expensive pig and scrap were always considered the best. Furnace control resided in one man who kept a watchful eye on the crucible and at the proper time bellowed, "It's melted; let's pour it."

The new philosophy presupposes that the melting operation is, to some extent, a refining process. For even with the best control of raw materials, an uncontrolled atmosphere can contribute hydrogen, oxygen, nitrogen, and water vapor. In addition, various trace elements and other gases which do not appear in the standard chemical analyses can have marked effects upon casting soundness and mechanical properties. On the other hand, there are many cases in which a product obtained from refined low-cost scrap is of better quality than that obtained with the high-cost raw materials formerly used.

These facts have led the responsible furnace operator to study metallurgical thermodynamics. From this relatively new science, he learns that metal solutions, say nickel dissolved in liquid iron at 3000°F (1649°C), behave a good deal like the aqueous solutions he studied in chemistry. For example, at these high temperatures, liquid metals have appreciable vapor pressures (liquid manganese evaporates like a puddle of water on a hot summer day), and the possibilities and characteristics of various high-temperature reactions can be estimated by means of the familiar concepts of equilibrium. Although much of this research has been limited to liquid steel and liquid copper for ingot-making, the subject deserves careful study by the metallurgists of the cast-metals industry. We shall review this work and its applications under four principal headings: gases in metals (this chapter), control of chemical composition (Chapter 11), selection and control of melting processes (Chapter 12), review of metallurgical calculations (Chapter 13*).

---

* Chapter 13 duplicates some of the previously treated subject matters, but for purposes of review it is convenient to summarize the typical calculations in one place. In this way, too, it is possible to avoid lengthy interruptions of the text of other chapters.

10-2 Gases in metals. The presence of gas in metal leads to faulty castings and hence is responsible for many foundry failures. On the other hand, controlled use of gas can be a valuable tool not only in the refining process, but also for attaining certain properties in the solid state, such as austenite stability and graphitizing rate. It is worth while, then, to consider in some detail the behavior of gases in metals. The effect of late additions to the melt, the so-called inoculating effect, is closely related to inclusions and gas content, and hence this subject is also included in this chapter.

Let us first catalogue the ways in which gases occur in castings. Gas is found in metals either because it has been mechanically entrapped, or because there has been a change in solubility or a chemical reaction in the liquid metal. Although the first of these alternatives is an important source of trouble, we may dismiss it as a physical problem of mold design. A proper, nonaspirating gating system in combination with a well-vented permeable mold will eliminate this source of gas.

We shall be concerned here only with gases dissolved in the liquid metal or formed by chemical reactions with or within the metal. We shall also investigate the reaction of metals with the mold materials which leads to the formation of gas, as well as the problem of gas control during melting. The discussion may be conveniently divided into these parts:

(a) *Simple gases:* apparently only solution and evolution of a single gas are involved.

(b) *Complex gases:* several elements are dissolved in the melt and react chemically during or near solidification to form gas. The role of "scavenging" elements to prevent such gas formation is also considered in this category.

(c) *Inoculating effects:* In many cases a ladle addition of a deoxidizer may have important additional effects upon the metal structure. These phenomena are therefore included in this chapter.

The simple gases of greatest importance are hydrogen and nitrogen. Hydrogen will be reviewed first and, because it is almost universally present in cast metal, it will be given a major share of attention. In each instance the source of gas and then its control will be considered.

10-3 Hydrogen. It is desirable to discuss first the solubility of hydrogen in solid metals, since it is the difference between liquid and solid solubilities at the solidification temperature that determines the amount of gas that may precipitate upon freezing. Smith [1] divides metals which dissolve hydrogen in appreciable amounts into two groups: (1) iron-type (endothermic) metals, and (2) palladium-type (exothermic) metals. These metals exhibit some periodic relationship, as indicated in Fig. 10-1. The group comprising iron, cobalt, nickel, and copper (right-hand side of

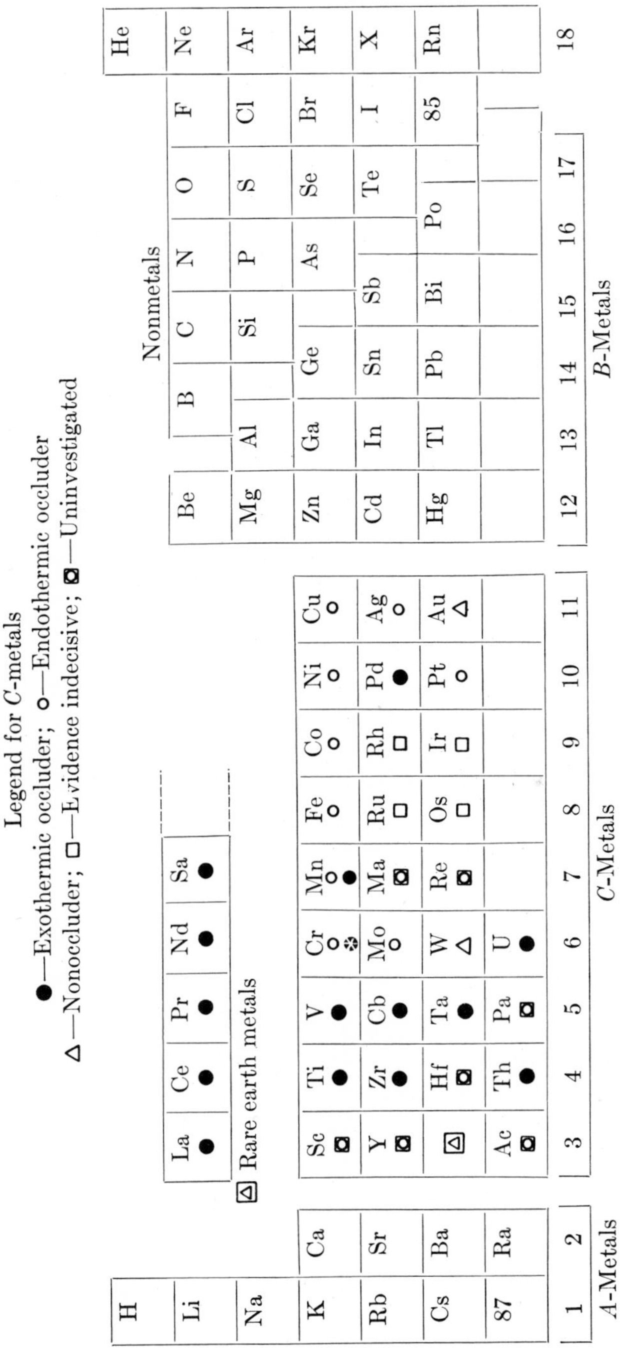

Legend for C-metals

●—Exothermic occluder; o—Endothermic occluder
●—Nonoccluder; □—Evidence indecisive; ◙—Uninvestigated

◙ Rare earth metals

Rare earth metals:

La ●	Ce ●	Pr ●	Nd ●	Sa ●

A-Metals:

1	2
H	
Li	
Na	
K	Ca
Rb	Sr
Cs	Ba
87	Ra

C-Metals:

3	4	5	6	7	8	9	10	11
Sc ◙	Ti ●	V ●	Cr o✿	Mn o●	Fe o	Co o	Ni o	Cu o
Y ◙	Zr ●	Cb ●	Mo o	Ma ◙	Ru □	Rh □	Pd ●	Ag o
◁	Hf ◙	Ta ●	W △	Re ◙	Os □	Ir □	Pt o	Au △
Ac ◙	Th ●	Pa ◙	U ●					

Nonmetals / B-Metals:

			B	C	N	O	F	He
Be								Ne
Mg	Al	Si		P	S		Cl	Ar
Zn	Ga	Ge		As	Se		Br	Kr
Cd	In	Sn	Sb		Te		I	X
Hg	Tl	Pb	Bi		Po		85	Rn
12	13	14	15	16	17	18		

B-Metals

FIG. 10–1. The solubility of hydrogen in various metals [1].

TABLE 10-1

SOLUBILITY OF HYDROGEN IN VARIOUS METALS
(Pressure = 1 atm)

Liquid solubility at solidus, cc/100 gm		Solid solubility, cc/100 gm	Difference, cc/100 gm
Mg	26.0	18.0	8.0
Al	0.7	0.04	0.66
Cu	5.5	2.0	3.5
Fe*	27.0	7.0	20.0

* For iron, convenient conversion factors are: 1 part per million (ppm = 0.0001% by weight) equals 1.11 cc/100 gm equals 0.0873 relative volumes (RV), that is, volumes H_2 per volume metal.

the table) is endothermic, while titanium, zirconium, and columbium (at the left) are exothermic; intermediate metals show both tendencies. The nature of the solution process, i.e., whether it is endothermic (heat absorbing) or exothermic (heat releasing), determines the amount of hydrogen dissolved. Endothermic metals absorb less hydrogen than exothermic metals do. (In the endothermic group, nickel, which dissolves a maximum of 1.56 volumes at 266°F (130°C), is the greatest absorber.) Also endothermic metals dissolve increasing amounts of hydrogen with increased temperature, whereas the reverse is true for the exothermic type. This observation agrees, of course, with the familiar LeChatelier principle: when a reaction liberates heat, the effect of increased temperature is to change the equilibrium point toward the direction of absorbing heat. The general opinion is that hydrogen forms hydrides or dissolves interstitially in the exothermic case (lattice distortion is noted) and merely dissolves, in defects in the metal lattice, in the endothermic reaction (no distortion). The solubility in either case can be represented by the equation

$$S = Ce^{-E_S/KT},$$

where S is the solubility, C and K are constants, E_S is the heat of solution of 1 mol of hydrogen, and T is the absolute temperature. A plot of $\log_e S$ versus $1/T$ is a straight line of slope $-E_S/K$. Hence, if E_S is positive (the solution is endothermic), the slope is negative, and solubility increases with temperature.

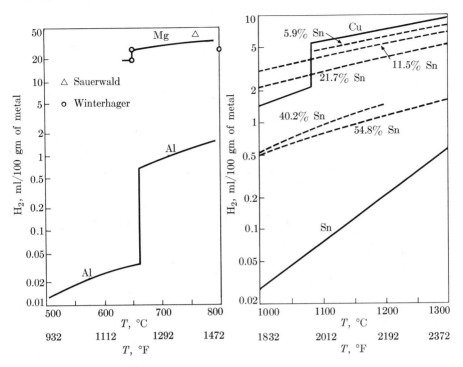

Fig. 10–2. Solubility of hydrogen at atmospheric pressure in aluminum and magnesium. (Data on aluminum from Ransley and Neufeld; data on magnesium from Sauerwald and from Winterhager [2].)

Fig. 10–3. Solubility of hydrogen at atmospheric pressure in copper, tin, and copper-tin alloys. (Data from Bever and Floe, and Sieverts [2].)

The importance of these relationships is twofold. First, if the solution is exothermic (gas solubility increasing with falling temperature), no trouble with gas precipitation during cooling should be expected, in contrast to the endothermic case. The common metals we shall deal with—aluminum, magnesium, copper, iron, and nickel—are all endothermic, while titanium and zirconium are exothermic. Solubility curves of hydrogen in various metals are shown in Figs. 10–2, 10–3, 10–4, and 10–5.

Although the amount of hydrogen by weight that can be dissolved appears very small, the volume *evolved* at solidification is great. The relative volumes for different metals are listed in Table 10–1.

The solubilities of hydrogen are usually given for a pressure of one atmosphere of pure hydrogen, and the question naturally arises as to the effect of low hydrogen pressures such as might prevail in the dissociation of water vapor from the combustion of furnace gases or from moist air.

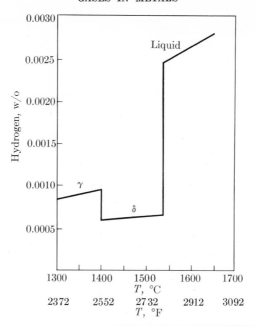

FIG. 10–4. The solubility of hydrogen in iron at atmospheric pressure [3].

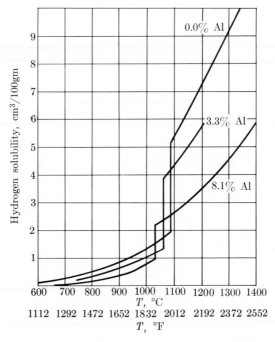

FIG. 10–5. Solubility of hydrogen in copper and copper-aluminum alloys. (Data from Röntgen and Moller [2].)

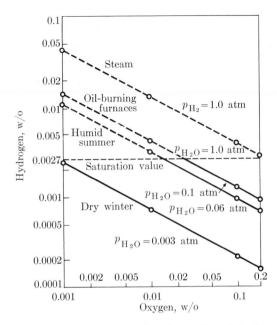

FIG. 10–6. Effect of deoxidation on the absorption of hydrogen from water vapor in liquid steel at 2190°F (1199°C) [3].

Sievert's law states that the amount of gas dissolved in the melt varies as the square root of the pressure of the gas in the atmosphere over the melt:

$$\% \, [\mathrm{H}]\dagger = K\sqrt{p_{\mathrm{H}_2}}.$$

We can evaluate K from the data for one atmosphere of pressure. For example, for magnesium,‡ $K = 26$ when $\%\,[\mathrm{H}]$ is expressed in cc/100 gm and p in atmospheres. If the pressure of hydrogen is reduced to 0.1 atm, the melt will contain 8.22 cc of dissolved hydrogen per 100 gm of metal. A special case develops when iron is covered with a slag layer which is permeable to water but not to hydrogen (Fig. 10–6). A well-deoxidized heat can dissolve more hydrogen from a humid summer atmosphere ($p_{\mathrm{H}_2\mathrm{O}} = 0.06$ atm) than from a hydrogen-rich atmosphere over a slag-free melt.

10–4 Removal of hydrogen. Since there is no simple dehydrogenating addition which can be used to eliminate hydrogen as a slag, care should be exercised to keep the hydrogen level to a minimum. Solid additions

---

† Brackets indicate that the element is dissolved in the metal.
‡ In the liquid state at the solidus temperature.

which form a purging gas in the melt are used, however. The solution of hydrogen is primarily due to the decomposition of water vapor which can come from damp furnaces, fluxes, scrap, air, ladles, and sand. Oil and grease are also sources of hydrogen.

Practically all methods for hydrogen removal center about Sievert's law, i.e., they seek to reduce the partial pressure of hydrogen over the melt. Obviously, inasmuch as a slag or oxide layer usually covers the melt, agitation is needed to reach equilibrium.

In the purging method usually employed for nonferrous metals, a dry insoluble gas such as chlorine, nitrogen, helium, or argon is bubbled through the melt. The hydrogen diffuses into the bubbles in an amount that satisfies the equation $\%[\mathrm{H}] = K\sqrt{p_{\mathrm{H_2}}}.$ In this way, the percentage of hydrogen is gradually reduced (see Chapter 13 for calculations). In nonferrous alloys with a high percentage of zinc, the high partial pressure of the zinc (over 1 atm) causes it to form bubbles in the melt that carry off hydrogen.

For steel and nickel-base alloys, carbon monoxide bubbles are used instead of nitrogen. This is done for two reasons. First, nitrogen is soluble in iron and hence cannot be regarded as an inert gas. Secondly, carbon is soluble in liquid iron and nickel and it is usually simpler to control the carbon content by lowering it to a given level by oxidation and subsequent recarburization. The oxidation of carbon, known as the carbon boil, is accomplished by adding oxygen in one of several ways. An air blast, an oxygen lance (oxygen under pressure blown through a pipe), or iron ore may be added. In each case CO bubbles form, and in each bubble the hydrogen tends to reach a partial pressure depending upon the hydrogen level in the liquid steel. Thus the CO bubbles flush out the hydrogen.

The data of Fig. 10–7 illustrate the effects that CO boil, furnace additions of ferroalloys, and tapping through the atmosphere have on the amount of dissolved hydrogen [12]. The data are derived from observations of acid and basic open hearths and basic electric furnaces. (The operation of these furnaces is described briefly in Chapter 12.) Notice that in all cases the hydrogen decreases during the carbon boil. After the oxidizing slag is removed or reduced by adding materials such as ferrosilicon (Fe-Si) or carbon, the hydrogen content increases, due to either a combination of hydrogen or water trapped in the alloys or to the absorption of hydrogen from the furnace atmosphere. The increase in hydrogen upon tapping and the effect of a wet ingot mold are also of interest. The reason that these small amounts of hydrogen in steel have received such intensive study is that even a low hydrogen level of 2 ppm can produce fissures (called flakes) in steel ingots. It is evident from the

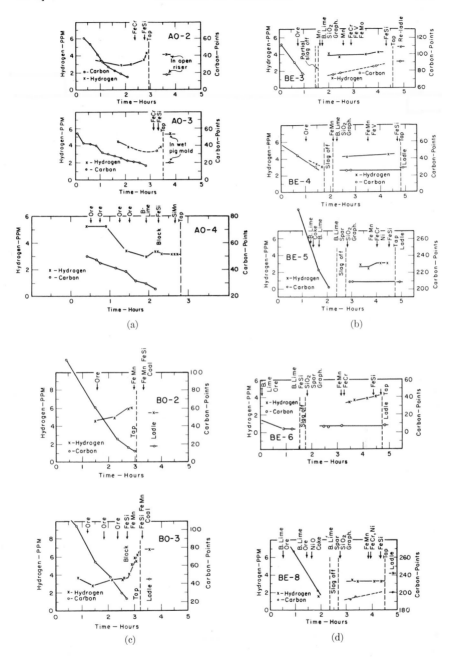

Fig. 10–7. (a) Logs of acid heats AO-2, AO-3, and AO-4. (b) Logs of basic electric heats BE-3, BE-4, and BE-5. (c) Logs of basic open hearth heats BO-2 and BO-3. (d) Logs of basic electric heats BE-6 and BE-8 [12].

data in Fig. 10–7 that it is difficult to reduce the hydrogen content to below 2 ppm by standard practices. To decrease the hydrogen level further, vacuum degassing is employed (see Section 10–9).

10–5 Nitrogen. While the solubility of hydrogen is of concern in the melting of both ferrous and nonferrous alloys, the solubility of nitrogen is significant only in the iron-base alloys and in the related iron-nickel-chromium-carbon alloys. Indeed, because of its insolubility, nitrogen is used as a flushing gas in aluminum and copper-base alloys. In many iron-base alloys, the nitrogen content is highly important not only from the standpoint of porosity but also because of its effect in solid solution or in the formation of nitrides. To illustrate these points, it may be recalled that many heat-resisting alloys require a stable austenitic structure. The following alloys have the same austenitic stability and illustrate the powerful effect of nitrogen in this connection.

	C	Cr	Ni	Si	N
Alloy 1:	0.15	25.0	20.0	2.0	0.0
Alloy 2:	0.15	22.0	13.0	1.4	0.20

In unalloyed steel the influence of nitrogen is often detrimental (Fig. 10–8). Nitrogen can combine with iron, aluminum, and other elements to form nitrides. Depending upon the type and time of nitride formation,

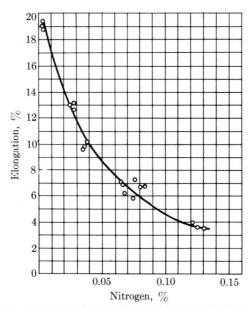

Fig. 10–8. Effect of nitrogen content on elongation of iron wire in tension. From N. Tschischewski, *J. Iron Steel Inst.* **92,** 47 (1915) [1].

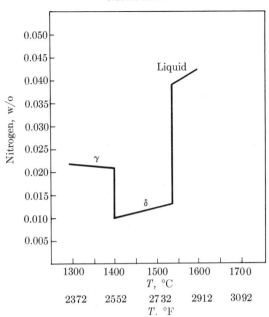

Fig. 10–9. Solubility of nitrogen in iron under atmospheric pressure [4].

the precipitate can either be of assistance in grain-size control or can lead to so-called "rock-candy" fractures which are characterized by low ductility and inferior impact strength. Discussion of the precipitation of nitride will be reserved until the solubility of aluminum nitride has been discussed, but from this brief comment it is evident that we should concern ourselves here with the control of nitrogen content. We will begin with the simple iron-nitrogen system and then discuss the more complex alloys.

The solubility of nitrogen in pure iron is plotted in Fig. 10–9. The solubility in a given phase follows Sievert's law, $\% \, [N] = K\sqrt{p_{N_2}}$. Just as in the case of hydrogen, there is a severe decrease in solubility as the metal passes from the liquid to the solid phase. Hence porosity may develop due to nitrogen evolution upon freezing. However, this happens relatively rarely because the carbon boil removes nitrogen as well as hydrogen, and thereafter nitrogen dissolves at a much slower rate than hydrogen because the passage through the slag retards the transfer of nitrogen from the air. It should be pointed out, however, that the presence of nitrogen compounds may lead to an increase in the nitrogen content of the bath, due to the greater solution potential of nitrogen in compound form. Thus, for example, calcium cyanamide, ammonia compounds, or even the nitrogen activated by the carbon arc may assume significance.

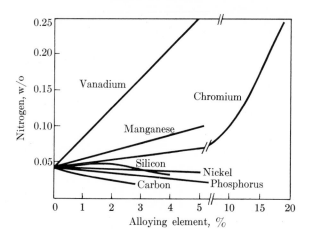

FIG. 10–10. Solubility of nitrogen in binary liquid iron alloys at 2190°F (1199°C) and atmospheric pressure [4].

From the standpoint of nitrogen solubility, engineering alloys may be grouped into two classes: (1) low-alloy steels and cast irons for which $\%N < 0.02$, and (2) highly alloyed steels, particularly those with high chromium content for which $\%N > 0.04$. Among the low-alloy steels, nitrogen content varies from 0.01 to 0.02% for Bessemer steel to 0.005 to 0.0015% for open-hearth and electric-furnace steels. These differences in nitrogen content are reflected in the steel's behavior during cold working and machining. The newer converter-type processes using oxygen have led to a reduction in nitrogen.

In the highly alloyed steels, nitrogen is often added deliberately in the form of 2% nitrogen ferrochromium. It is important to estimate the solubility of nitrogen in these complex materials. The separate effects of various alloying elements upon nitrogen solubility is shown in Fig. 10–10. To determine the solubility in a complex alloy, the following method based upon the activity coefficients has been developed [13, 14].

We have seen that the solubility is proportional to the square root of the pressure which in turn is related to the activity (see Chapter 13). For the ranges of solubility in question, we can approximate the effect of a third element A upon the solubility of nitrogen in iron by the activity coefficient of nitrogen in the ternary alloy for low concentrations:

$$f_N^A = \frac{\%[N°] \text{ (dissolved in pure iron at 1 atm pressure } N_2)}{\%[N] \text{ (dissolved in ternary alloy at 1 atm pressure } N_2)},$$

where f_N^A is the activity coefficient of nitrogen in the Fe-N-A system. To

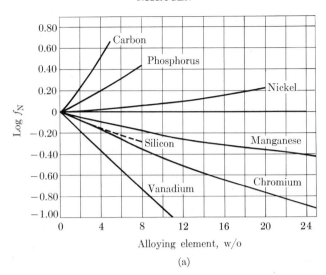

(a)

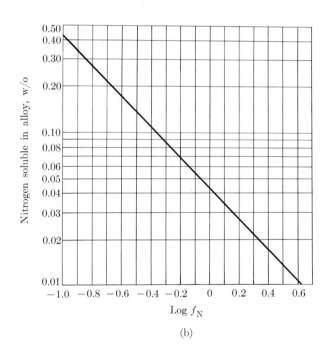

(b)

FIG. 10–11. (a) Activity coefficients of nitrogen at 1600°C (2912°F) as a function of alloy content. Graph is used to compute log f_N in an alloy steel. (b) Chart for converting log f_N into w/o nitrogen soluble in an alloy at 1600°C (2912°F), under a pressure of 1 atm N. Log f_N was determined from Fig. 10–1.

determine the combined effects of several elements A, B, C, we then employ the relation

$$\log f_N^{A,B,C} = \log f_N^A + \log f_N^B + \log f_N^C .$$

Typical values may be obtained from Fig. 10–11(a) for a temperature of 2912°F (1600°C).

Since f_N is now known for the complex alloy, and $\%[N°]$ at 2912°F is 0.043, $\%[N]$ in the complex alloy may be calculated. This is done more conveniently by means of the graph of Fig. 10–11(b). To present a typical calculation, we shall determine the amount of nitrogen soluble at a pressure of 1 atm N_2 in a type 308 stainless steel containing 0.40% C, 2.0% Mn, 2.0% Si, 20% Cr, and 10% Ni. From the algebraic sum of the log f_N's in Fig. 10–11(a), we obtain a value of

$$+0.09 -0.05 -0.04 -0.76 +0.08 = -0.68.$$

From Fig. 10–11(b), we then find a solubility of 0.20% N, or five times that existing in pure iron. A more refined method of calculation is given in reference 13.

The importance of this value for maximum solubility under a pressure of 1 atm N_2 is that it represents the maximum amount of nitrogen which can be introduced and retained in a melt under a neutral slag. If this value is exceeded, as during cooling, nitrogen bubbles can form and reduce the level rapidly to the equilibrium point. On the other hand, if the quantity of dissolved nitrogen is below this maximum, the approach to equilibrium with the partial pressure of 0.8 atm N_2 in the normal atmosphere of air will be slow because it will depend on the diffusion of nitrogen through the slag layer.

Unfortunately the solubility of nitrogen in the solid state in the complex alloys is not known. However, if maximum solubility in the liquid is reached, there is a strong probability of porosity in the ingot. Fortunately, there are several factors that may prevent formation of nitrogen bubbles. First, if the phase diagram has been altered by the alloying element so that austenite (γ) instead of δ-ferrite is formed upon freezing, the nitrogen solubility may be much greater (extrapolate the γ-line of the Fe-N phase diagram). Secondly, the nitrogen solubility in the solid state may be enhanced by the alloying element. Finally, the nitrogen may precipitate as nitride rather than as N_2 bubbles, particularly if Zr or Ti is present. The solubility products for various elements at 2912°F (1600°C) are listed in Table 10–2.

The table represents the combinations of concentrations at which a particular nitride will precipitate at 2912°F. If we consider the effect of 0.10% Ti, we see that the amount of nitrogen remaining in solution could

TABLE 10–2

DENITRIFYING POWER OF CERTAIN ELEMENTS
IN LIQUID IRON [4]

Element	Constant	Value at 2912°F (1600°C)
Al	$\%[\text{Al}] \times$ w/o N	0.55
Si	$\%[\text{Si}^{3/4}] \times$ w/o N	14.0
Ti	$\%[\text{Ti}] \times$ w/o N	0.00014
V	$\%[\text{V}] \times$ w/o N	1.5
B	$\%[\text{B}] \times$ w/o N	0.55
Zr	$\%[\text{Zr}] \times$ w/o N	0.00014

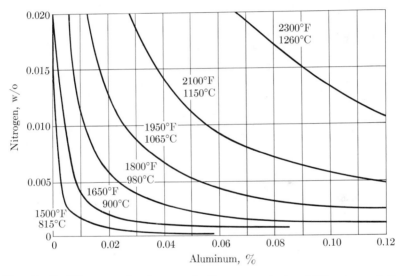

FIG. 10–12. Minimum aluminum and nitrogen contents yielding a precipitate of AlN at various temperatures. Contents of aluminum and nitrogen above or to the right of any isotherm are precipitated.

be 0.00014/0.1, or 0.0014%. Although much higher aluminum contents would be required at this temperature to reduce the nitrogen to this level, it is important to note that the solubility product decreases with decreasing temperature. Upon cooling in the solid state, the formation of aluminum nitride, particularly at grain boundaries, may have two effects. The nitride may inhibit grain growth and lead to *fine-grained* steel. In large quantities, however, the aluminum nitride precipitate may be excessive and lead to grain-boundary fracture in cast steel. The variation in the constant for the reaction $\%[Al] \times \%[N] = K$ as a function of temperature is given by the equation:

$$\log K = -\frac{7400}{T} + 1.95,$$

where T is in degrees absolute. The graph of this equation is plotted in Fig. 10–12. Note that an aluminum content of 0.10% in solution accompanied by a nitrogen content over 0.013% would lead to precipitation of aluminum nitride on cooling to about 2300°F (1260°C).

10–6 Complex gases. CO in steel. Let us now turn to the somewhat more complex generation of gas during solidification, in which one constituent is rejected and reacts with a second which is still in solution. The two principal cases are the generation of CO in steel (and nickel alloys) and the complex $H-O-H_2O-S-SO_2$ equilibria in copper. No complex gases are encountered in aluminum and magnesium alloys.

The reaction of carbon with oxygen is one of the most important reactions in liquid metal. We have already discussed how the CO boil in a heat can be used to advantage to reduce the hydrogen and nitrogen contents, but there are obvious disadvantages to the same boil during solidification.* Here the tendency to produce $CO + CO_2$ is even greater than in liquid steel, because as steel solidifies, the dissolved oxygen is rejected as FeO, which reacts with the dissolved carbon to form CO and CO_2.

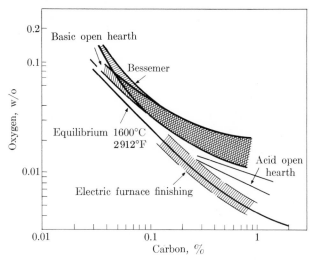

Fɪɢ. 10–13. Oxygen contents in commercial liquid steel for various steelmaking practices [8].

---

* In castings. The CO "rimming action" is present in rimmed steel ingots.

Before we examine the means at our disposal for preventing this gas evolution, let us study the carbon-oxygen equilibrium line in Fig. 10–13. Pure iron dissolves a maximum of 0.21% O at 2912°F (1600°C), and carbon reduces the solubility by the reaction $C + O = CO$ (also CO_2). Of course, if the CO is swept away, the reaction continues to lower the carbon content just as in the familiar open-hearth, converter, or electric-furnace oxidizing practices. In the commercial furnace practices illustrated in Fig. 10–13, the oxygen is higher than called for by equilibrium because the carbon content of the heat is being reduced, and an oxygen differential in the slag is needed. The samples from the electric furnace heats were taken under a reducing finishing slag, and so conform more closely to the indicated equilibrium conditions.

The calculations for equilibrium amounts of C, CO, CO_2, and dissolved oxygen are rather involved. It would be simple if only CO were formed, since we would then have the equilibrium

$$[C] + [O] = CO \quad \text{(gas)}, \qquad K = \frac{p_{CO}}{a_C \cdot a_O} \, ;$$

or assuming that the activity of each element is equal to the proportion (in per cent) in which it is present, and using the special symbol m, we have the relation

$$m = \frac{\%[C] \times \%[O]}{p_{CO}} \, .$$

Unfortunately, neither of the above conditions holds: CO_2 is always present, and the activity of a gas does not always equal the percentage in which it is present. Correction factors have been developed [4] to aid in such computations, or the values can be estimated from Fig. 10–13. Let us assume that 0.01% oxygen is present in a 0.3% carbon steel. If the steel is cooled quickly to the solidification range, 0.018% CO can be formed (neglecting CO_2) since the solubility of oxygen or FeO in solid iron is very slight. This corresponds to a volume of 91 cc per 100 grams of iron at 2732°F (1500°C), or to a volume ratio of over 6 to 1. (These values are based on the assumption that the conversion of dissolved O to CO is complete; actually, some FeO would precipitate and remain in the form of inclusions.)

We may now study the role of deoxidation, which in all cases consists of the addition of an element with a smaller deoxidation constant than that of the C-O equilibrium. Typical values are contained in the summary given in Fig. 10–14 [4]. The addition of 0.5% silicon to the 0.3 carbon steel just mentioned reduces the [O]-value to 0.009%, a very slight reduction from 0.010% at 2910°F (1599°C). However, at lower temperatures, e.g., the freezing range, silicon becomes more potent. Manganese has little to offer when added alone. However, even small quan-

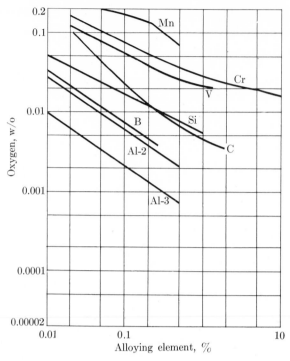

FIG. 10–14. Summary of information on deoxidation (2912°F, 1600°C) [4]. (Note discrepancy in case of aluminum: Al-2, Hilty and Crafts; Al-3, Wentrup and Hieber.)

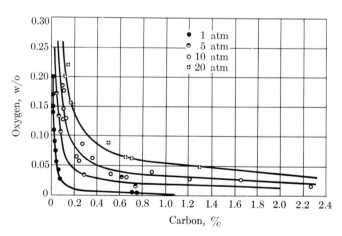

FIG. 10–15. Effect of carbon on the oxygen content of the melt under equilibrium conditions, for various atmospheric pressures and at 1540°C (2804°F). (After Marshall and Chipman [4].)

tities (0.1%) of the more powerful deoxidizers, such as Ti, Zr, or Al, can reduce the dissolved oxygen by several orders of magnitude. Also, residual amounts of these alloying elements react rapidly with any FeO that is formed during cooling to the solid state, thus providing a buffering action which prevents the formation of CO. The quantitative variation of the deoxidation constant with temperature is discussed in Chapters 11 and 13. In passing, it should be mentioned that carbon itself becomes a better deoxidizer with reduced pressure. This effect is employed in vacuum melting and vacuum degassing since the oxygen can be reduced to very low values in this manner and the solid inclusions which accompany aluminum deoxidation are avoided. The variation in [O] level with pressure is shown in Fig. 10–15.

10–7 Complex gases in copper and copper alloys. Hydrogen, sulfur, carbon, and oxygen are less soluble in solid copper than in liquid copper. During solidification, these gaseous elements therefore become concentrated in the liquid copper, and bubbles may form if the liquid solubility is exceeded *at any time* and the sum of the partial pressures of the gases exceeds one atmosphere. We shall see shortly that the danger of porosity presented by the compounds of these gaseous elements is greater than that deriving from the separation of the elements themselves.

Before we calculate the formation of these compounds, let us establish a basis for our computations by examining (1) the Cu-H, Cu-S, and Cu-O phase diagrams of Figs. 10–16 and 10–17 (the Cu-C data of reference 9 are also of interest in considering the possibility of CO formation), (2) the equilibrium constants for the reactions of these gases in liquid copper to form H_2O and SO_2, and (3) the partial pressures of H_2O, SO_2, and CO_2 in commercial furnace atmospheres.

EQUILIBRIUM CONSTANTS

Hydrogen: $\quad$ Cu (l) + H_2O (g) = [O] in Cu + [2H] in Cu,

$$[H] = 3.04 \times 10^{-7} \sqrt{p_{H_2O}/[O]} \quad \text{at } 1083°C \ (1981°F),$$

$$[H] = 7.96 \times 10^{-7} \sqrt{p_{H_2O}/[O]} \quad \text{at } 1250°C \ (2282°F).$$

Sulfur: $\quad$ Cu (l) + SO_2 (g) = [S] in Cu + [2O] in Cu,

$$p_{SO_2} = \frac{[S] \times [O^2]}{3.31 \times 10^{-5}} \quad \text{at } 1250°C,$$

$$p_{SO_2} = \frac{[S] \times [O^2]}{0.98 \times 10^{-5}} \quad \text{at } 1083°C.$$

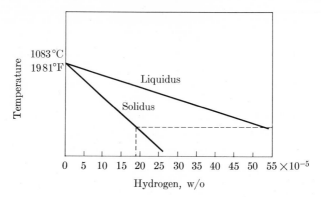

FIG. 10–16. Schematic copper-hydrogen constitution diagram [10].

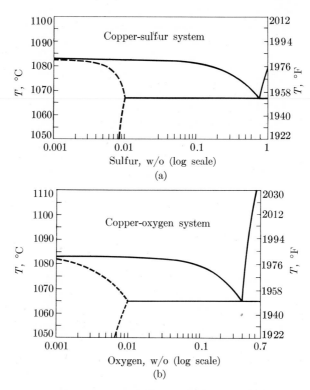

FIG. 10–17. (a) Constitution diagram of the copper-sulfur system [10].
(b) Constitution diagram of the copper-oxygen system [10].

$$\text{Cu(l)}+H_2O(g) \rightleftharpoons Cu_2O(\text{in Cu})+H_2(\text{in Cu})$$

$$[H] = 7.96 \times 10^{-7} \sqrt{p_{H_2O}/[O]}, \text{ at } 1250°C$$

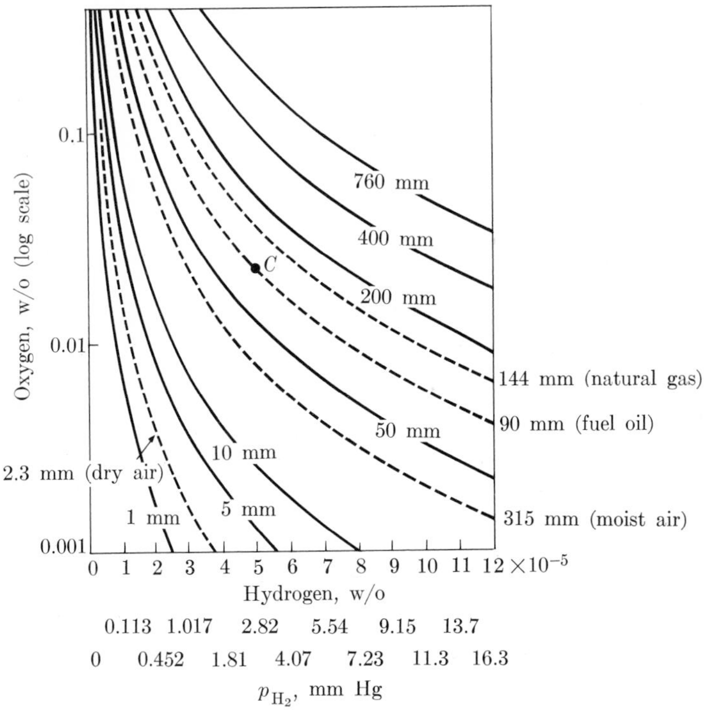

FIG. 10–18. Equilibrium of water with molten copper at 1250°C (2282°F) [10].

Copper-H_2O problem. The plot of Fig. 10–18 illustrates that the combined effect of oxygen and hydrogen is far more severe than that of hydrogen alone. From the data for hydrogen solubility previously given and from Sievert's law, we calculate that the hydrogen partial pressure in the atmosphere above the melt at 1083°C would have to exceed 0.125 atm (97 mm) to cause porosity on *solidification.* (From Fig. 10–3 solid solubility = 2 ml/100 gm; 0.132 atm H_2 produces this value in liquid.*)

If both oxygen and hydrogen are present at freezing, as shown in Fig. 10–19, a partial pressure of less than 0.00134 atm H_2 or a hydrogen content of $2 \times 10^{-5}\%$ by weight is sufficient to produce a total pressure of 760 mm H_2O at 0.39% [O] (point B of Fig. 10–19). It may be contended that this is a very high oxygen level, but the Cu-O phase diagram shows that if 0.01% [O] is present, this quantity will increase to 0.39% during

---

* K (at 1083°C) = 5.5; %[H] = $K\sqrt{p_{H_2}}$; $p_{H_2} = 2^2/5.5^2 = 0.132$ atm.

$$Cu(l)+H_2O(g) \rightleftharpoons Cu_2O(in\ Cu)+H_2(in\ Cu)$$

$$[H]=3.04\times10^{-7}\sqrt{p_{H_2O}/[O]},\ at\ 1083°C$$

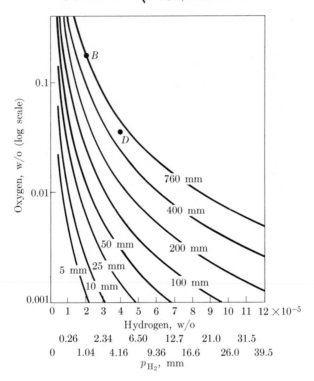

FIG. 10–19. Equilibrium of water with molten copper at 1083°C (1981°F) [10].

freezing.* Probably even less oxygen is needed in the melt to produce this eutectic oxygen content, since anything short of complete diffusion in the solid would lead to higher oxygen contents in the liquid.

The curves in Fig. 10–18 represent the various quantities of oxygen and hydrogen dissolved in copper at equilibrium with different furnace atmospheres. Let us follow the gas evolution in liquid copper as developed in Fig. 10–18 [10]. At 1250°C (2282°F) after poling in an oil-fired refining furnace, about 0.02% [O], and hence $5.4\times10^{-5}\%$ [H] (point C of Fig. 10–18) remain in the melt. We know from experience that the level of dissolved oxygen reaches 0.035% by the time molds are poured, and % [H] is then about 4×10^{-5} (point D of Fig. 10–19). If these percentages are retained by rapid cooling to the freezing range (1083°C or 1981°F), then a partial pressure of about 600 mm H_2O results (see Fig. 10–19), and hence no gas evolution takes place *at the start* of freezing. The copper liquid will, however, become enriched in [O] and

---

* O-content of residual liquid follows liquidus to 0.39% [O].

$$Cu + H_2O \rightleftharpoons Cu_2O + H_2$$
$$p_{H_2} + p_{H_2O} = [H]^2 \times 10^{14}(0.113 + 159.5[O])$$
at 1250°C

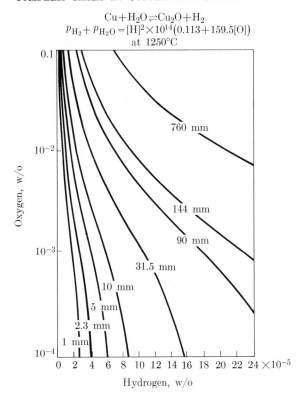

FIG. 10–20. Equilibrium of water and hydrogen with molten copper at 1250°C (2282°F) [10].

[H], according to the phase diagrams, and it can be calculated that after about 10% of the metal solidifies, H_2O will be evolved until solidification is complete. At moderate oxygen and hydrogen levels, porosity is therefore readily encountered.

Let us now consider deoxidized copper and its alloys, where very low oxygen contents are obtained. The partial pressure of hydrogen again becomes important and must be added to that of H_2O. Figure 10–20 illustrates the combined partial pressures of H_2 and H_2O in equilibrium with [O] and [H] at melting temperatures in different fuel atmospheres. Figure 10–21 shows the [O-H] balance at freezing that will exceed atmospheric pressure ($H_2 + H_2O$) and therefore lead to bubble formation.

In these low-oxygen alloys, as in the poled copper previously discussed, it is necessary to consider the partial pressure in the last liquid to solidify. Curves (2) and (1) show the relation between the gas pressures at the start and the end of solidification. Any combination of [H] and [O] which lies on curve (2) will reach a final pressure of 760 mm. The dashed

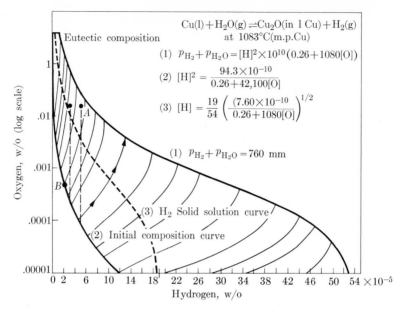

$Cu(l) + H_2O(g) \rightleftharpoons Cu_2O(\text{in } l \text{ Cu}) + H_2(g)$

at $1083°C$(m.p.Cu)

(1) $p_{H_2} + p_{H_2O} = [H]^2 \times 10^{10}(0.26 + 1080[O])$

(2) $[H]^2 = \dfrac{94.3 \times 10^{-10}}{0.26 + 42,100[O]}$

(3) $[H] = \dfrac{19}{54}\left(\dfrac{7.60 \times 10^{-10}}{0.26 + 1080[O]}\right)^{1/2}$

(1) $p_{H_2} + p_{H_2O} = 760$ mm

(3) H_2 Solid solution curve

(2) Initial composition curve

Eutectic composition

Oxygen, w/o (log scale)

Hydrogen, w/o

Fig. 10–21. Equilibrium of water and hydrogen with molten copper at 1083°C (1981°F) [10].

line, curve (3), indicates the maximum [H] which can be held in the solid in equilibrium with the liquid of curve (1). Also, any metal composition above curve (2) will exhibit porosity, and this will occur before the final solidification. For example, consider the composition given by the point A at 0.015% [O] and 5×10^{-5}% [H] (Fig. 10–21). This is related to a composition of 0.00075% [O] and 1.8×10^{-5}% [H] in that the liquid of the latter composition (point B) at start of freezing will pass by the point of 0.015% [O] and 5×10^{-5}% [H] during solidification. Material whose initial [H-O] balance is higher will show gas sooner, and the liquid will progress upward in the [H-O] balance on curve (3) while evolving gas.

Another example is shown by the arrows. In metal with an initial composition of 0.0001% [O] and 4.6×10^{-5}% [H], the dissolved oxygen and hydrogen will concentrate to 0.00385% [O] and 13×10^{-5}% [H] in the last liquid to freeze.

To avoid gas, then, it appears that the starting [H-O] combination must lie below curve (2). However, when a deoxidizer such as phosphorus is used, residual amounts will prevent buildup of [O] in the liquid because the deoxidation product will still hold for the remaining liquid. The residual P therefore acts as a buffer. As an example, if we start with 5.0×10^{-5}% [H] and 0.02% [P], the oxygen will be buffered at 0.002%.

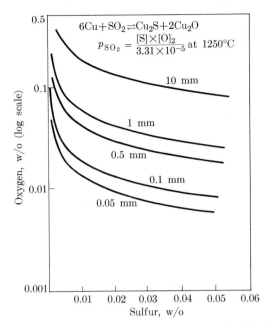

FIG. 10–22. SO$_2$ equilibrium with molten copper at 1250°C (2282°F) [10].

Since more hydrogen than the amount given can be present in the solid before $p_{H_2} + p_{H_2O}$ exceeds 760 mm, no bubbles will form (curve 3).

The advantage of starting with lower hydrogen content, as in electric-furnace melting, can also be demonstrated for high-conductivity copper, where a minimum of deoxidizer is desired. If [H] at the start can be held to $3 \times 10^{-5}\%$ and if [O] is maintained at, say, 0.0095%, the pressure can be kept below 760 mm. For this a residual P of 0.005% is sufficient, compared with the 0.02% P required for the reverberatory-furnace metal containing a higher percentage of dissolved hydrogen.

In copper-tin alloys, the solubility for hydrogen is lower (Fig. 10–3), and less deoxidation is needed. The copper-zinc alloys have a very low oxygen content, and the zinc boil tends to lower the hydrogen content. These materials are therefore relatively free of H$_2$O formation, and hydrogen is the principal offender due to the deoxidizing effect of the zinc.

The SO$_2$ problem. A similar analysis has been made for the Cu-S-SO$_2$ equilibrium [10]. The S-O-p_{SO_2} chart based on the results obtained (Fig. 10–22) apparently indicates seemingly low values of p_{SO_2}. However, just as in the case of the H-O equilibrium, the segregation effect during freezing, as well as the change of the equilibrium constant with temperature, can lead to high pressures at freezing, as shown in Fig. 10–23. The curve labeled $p_{SO_2} = 760$ mm represents the [O]- and [S]-values neces-

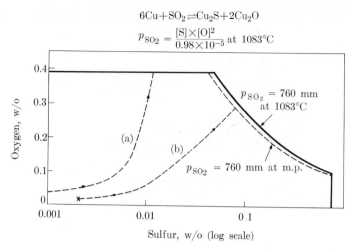

$$6Cu + SO_2 \rightleftharpoons Cu_2S + 2Cu_2O$$

$$p_{SO_2} = \frac{[S] \times [O]^2}{0.98 \times 10^{-5}} \text{ at } 1083°C$$

Fig. 10–23. SO_2 equilibrium with molten copper at the melting point [10].

sary in liquid copper at 1083°C (1981°F) to reach 1-atm pressure. At the copper-oxygen eutectic ([O] = 0.39%), at least 0.4% [S] is needed, and if the percentage of [O] is lower, the level of [S] must be correspondingly higher.

If we compute the increase in [O] as the liquid freezes (from the Cu-O diagram), we find that, starting with 0.02% [O], the last 2.5% of liquid is of eutectic composition (0.39% [O]). Similarly, from the Cu-S diagram, starting with 0.003% [S] and assuming complete diffusion, we see that when only 2.5% liquid is left, the level of [S] will have increased to 0.04%. A typical combination of oxygen and sulfur levels is illustrated by curve (b) in Fig. 10–23, which shows that a melt starting with 0.002% [S] and 0.012% [O] can reach a pressure of 760 mm with 0.09% [S] and 0.3% [O] in the remaining liquid.

These data indicate that the sulfur dissolved from fuel can produce gas. To avoid this effect, the oxygen content should be kept high to reduce [S] at a given p_{SO_2} during the early stages of the melt (Fig. 10–23). Once the metal has been deoxidized, minimum contact with SO_2 is essential to prevent S pickup to the new O-S equilibrium value of the deoxidized metal.

10–8 General summary of gas control. Despite the wide difference in degrees of reactivity, melting points, and types of furnaces used, a general common pattern for avoiding gas porosity in castings emerges from the discussion just completed. In almost all cases it is advisable to melt under dry oxidizing conditions which result in a boil, or to purge with a neutral gas. This helps to eliminate hydrogen and, in the case of steel,

TABLE 10-3

PRINCIPAL SOURCES OF GAS POROSITY

Alloy group	H	H_2O	CO	Other
Magnesium base	1			
Aluminum base	1			
Copper base	2	1	3	SO_2 [1]
Nickel base	2	2	1	
Iron and low-carbon low-alloy steel	2		1	
High-alloy steel	1			
Cast iron	2			N
Magnesium-treated cast iron	1			
Titanium and zirconium	No gas porosity (exothermic occluders)			

nitrogen. Where oxidation has left dissolved oxygen in the melt, as in copper, nickel, and steel, it must be removed with a deoxidizer. From this point onward, the melt is susceptible to hydrogen solution from humid air, furnace atmospheres, damp refractories, and the mold, and must be handled with special care.

The principal sources of gas *porosity* for the different alloys are summarized in Table 10-3, where the severity of the effect is denoted by 1, 2, and 3, with 1 being the most severe.

10-9 Vacuum melting. At the present time, vacuum melting, stream degassing, as well as the use of controlled atmospheres and reduced pressures, are receiving increasing attention, and it is frequently claimed that these processes are unusually effective.

Since the relative merit of starting with pure melting stock versus relying on the effect of the vacuum-melting technique *per se* is still a matter of controversy, no description of unique properties will be attempted at this time. However, it is profitable to consider the principles and basic advantages of these relatively new methods.

Vacuum melting is mainly used for the following purposes:

(1) To prevent the combination of reactive elements in the melt with the normal atmosphere and hence the formation of oxides and nitrides, which leads to a drossy melt, inclusions, and poor surface quality.

(2) To prevent the solution of gases such as H, N, SO_2 in the melt or to remove dissolved gases from the melt. Vacuum melting itself will not eliminate dissolved oxygen in steel or in copper; in fact, with high-

vacuum techniques, the dissolved oxygen may be increased as the dissociation pressure of metal oxides in the melt or in the refractory is approached.

As an example of gas removal, let us consider hydrogen dissolved in aluminum. Pure aluminum dissolves about 0.8 ml of hydrogen per 100 gm of metal in liquid at the freezing point, and 0.04 ml of hydrogen per 100 gm of metal in the solid (Fig. 10–2). As a melt containing 0.4 ml of hydrogen per 100 gm of metal freezes, at just over 50% solidification, the hydrogen content in the melt will be greater than the quantity which is in equilibrium with 1 atm of H_2. At this point, hydrogen bubbles can form on foreign nuclei, and some are trapped, leading to unsoundness. It should be noted that so far as bubble formation in freezing metal is concerned, the composition of the gas over the melt is not important—it is the total pressure over the melt which is significant. For example, there is a common test for gassiness in which a sample of liquid is placed in a low-pressure chamber to solidify. In a crucible of metal tested in this manner the melt will exhibit greater bubbling than if allowed to solidify at 1 atm. These bubbles occur when the quantity of hydrogen in the melt exceeds that which would be in equilibrium with a hydrogen pressure corresponding to the chamber pressure. This observation can be verified by means of Sievert's law.

As a corollary, we note that simply reducing the pressure will also lead to bubbling, even in the absence of partial freezing. Consider again the aluminum melt containing 0.4 ml of hydrogen per 100 gm of metal, but now at 100°F (56°C) above the freezing point. At this temperature, 0.9 ml H_2/100 gm of metal represents the equilibrium solubility for 1 atm of hydrogen. Note that in this case bubbling will take place when the pressure is lowered, whether or not hydrogen happens to be in the atmosphere over the melt.

Now let us calculate the reduction in pressure required to obtain bubbles in the melt containing 0.4 ml H_2/100 gm of metal. For the temperature given (100°F above the freezing point),

$$K = \frac{\%H^2}{p_{H_2}} = \frac{0.9^2}{1} = 0.81.$$

Let p_x be the pressure of H_2 in equilibrium at 0.4 ml/100 gm, or

$$p_x = \frac{(0.4)^2}{0.81} = 0.2 \text{ atm.}$$

Therefore, as the pressure over the melt is reduced below 0.2 atm, hydrogen bubbles will tend to form.

The question may now be asked, how far should the pressure be reduced to degas the metal sufficiently so that bubbles will not form when the melt is poured at atmospheric pressure? The solubility of the gas in the solid is 0.04%. Hence, assuming that this solubility is the same throughout the solid metal at the freezing temperature, the melt should be degassed below 0.04%, to avoid gas evolution upon freezing. We may calculate now that p_x, the pressure of hydrogen over the melt in equilibrium with 0.04% [H] is 0.002 atm or 1.52 mm Hg. If we then evacuate to 0.002 atm, the dissolved hydrogen will be 0.04%. If we now raise the pressure over the melt to 1 atm during pouring, no gas will be evolved. This presupposes that no solution of hydrogen takes place while the pressure over the melt is being raised. (Dry air or an inert gas can be used.) In other words, evacuating *and pouring* at reduced pressure do not eliminate porosity, but the pressure must be raised before pouring.

A practical application of this method, which greatly improves the quality of melts, has been developed for copper-base alloys [15]. After melting, the crucible is placed under vacuum for five minutes, and the dissolved gases literally bubble out.

Other work has shown that similar effects may be obtained with certain nickel-base alloys which are susceptible to porosity.

10–10 Inclusions. Since the type and manner of gas elimination can have a strong effect upon inclusions in the casting, these phenomena are best discussed here while the degassing treatments are still in mind.

Inclusions in steel. A major share of the investigations has focussed on steel inclusions, and various inclusions have been identified—oxides, silicates, nitrides, and sulfides; however, it has been found that only sulfides have a pronounced effect upon mechanical properties. These sulfide inclusions, classified as types I, II, and III, respectively, are illustrated in Fig. 10–24. As expected, the best ductility is obtained with the random globular inclusions, type I, and the poorest with the grain-boundary or eutectic shape, type II (Fig. 10–25). Type III provides slightly less ductility than type I. The effects of various amounts of deoxidizers are illustrated in Fig. 10–26.

The question may be asked: "If inclusions of type I can be obtained without aluminum additions, why is aluminum used?" The answer is given in the preceding section, in which the requirements for avoiding gas porosity are discussed in detail. The transition from globular, to eutectic, to the more random type of sulfide distribution (types I, II, and III) with increasing aluminum is explained as follows. In type I (no aluminum addition), a higher oxygen level exists, and iron oxysulfides with a high melting point are formed. These do not, therefore, occur at grain boundaries, but rather in random distribution. A small amount of aluminum lowers the level of dissolved oxygen, and the oxysulfides are replaced by low-melting sulfides at the grain boundaries (type II). The

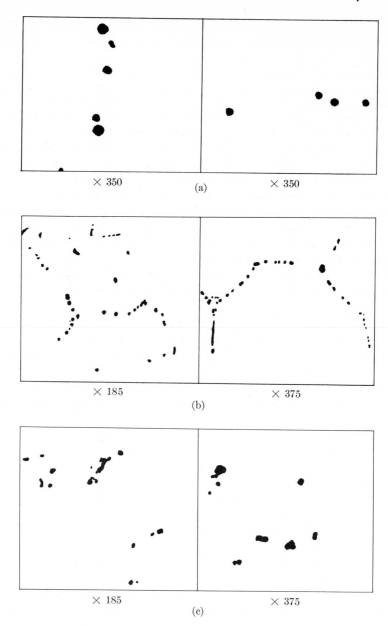

FIG. 10–24. (a) Globular silicate and sulfide inclusions. No addition of aluminum. Type I, high ductility, unetched (× 350) (Sims and Dahle). (b) Eutectic sulfide inclusions with occasional clusters of alumina. 0.025% aluminum added. Type II, low ductility, unetched (Sims and Dahle). (c) Duplex sulfides, large and irregular constituent Al_2S_3. Some alumina present. Type III, ductility good, unetched (Sims and Dahle).

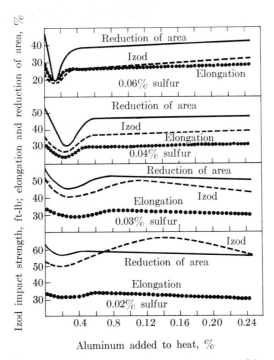

Fig. 10–25. Effect of aluminum and sulfur on ductility and impact strength of normalized medium-carbon cast steel. (Sims and Dahle)

addition of more aluminum results in the formation of aluminum sulfides which have a higher melting point than type II but are not so randomly distributed as type I.

This elementary explanation requires further elucidation by means of the concept of interfacial tension [6]. If the surface tension of a second phase is about equal to that of the matrix, the phase will take a triangular shape with dihedral angles of 120°. If the surface tension of the second phase is greater than that of the matrix, the included angle will be more than 120°, and a globular shape will be approached. If the surface tension of the second phase is smaller than that of the matrix, the phase will tend to spread out along the grain boundaries and finally assume the form of thin films. This theory, which deserves further attention, may replace the somewhat oversimplified and naive freezing-point theory, which is based on weak experimental data. The reversal of the behavior of some elements, as shown in Fig. 10–26, may also be considered from this standpoint.

It should also be emphasized that the above inclusion studies were all performed upon 1-inch test bars. Recent investigations indicate that in commercial castings, particularly of heavy sections, a high aluminum content is not sufficient to prevent inclusions of type II.

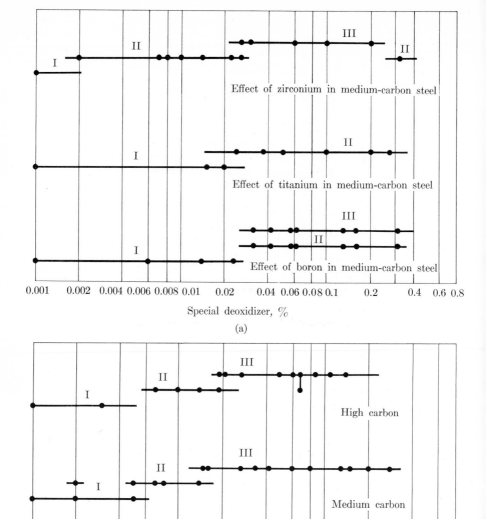

Fig. 10–26. (a) Effect of varying amounts of Zr, Ti, B on type of sulfide inclusions in medium-carbon steel [5]. (b) Relation of carbon content to effect of aluminum on the type of sulfide inclusions in cast steels [5].

10–11 Inoculating effects. Ladle additions. Ladle additions and so-called inoculations have been the most carefully guarded secrets in the whole field of melting. Many important variations in engineering properties are obtained by adding a given amount of alloy to the ladle, rather than directly to the furnace. For example, in gray cast iron there is a delicate balance between the reactions,

(1) liquid → iron carbide + austenite, and

(2) liquid → graphite + austenite,

either of which may occur on freezing. If reaction 1 occurs, the resulting product is hard and unmachinable, whereas the structure resulting from reaction 2 possesses excellent machinability.

The carbide-forming tendency can be greatly reduced by the addition to the ladle of 0.10 to 0.50% silicon in the form of ferrosilicon (75 to 85% silicon), calcium silicide, or one of a variety of special silicon alloys. The silicon added to the ladle is far more potent in producing graphitization than a similar amount melted down with the charge (even if oxidation during melting is accounted for). Many theories have been set forth to explain this effect. The two principal alternatives suggested are (a) that the high, although short-lived, silicon content around the pieces of ferrosilicon in the melt produces *graphite* nuclei for reaction 2, and (b) that the silicon addition produces a host of submicroscopic silicate nuclei for reaction 2.

For steel, the procedure of adding aluminum to the ladle to produce fine-grained steel is well known. If the same amount of aluminum were added earlier in the melt, the inclusions would have time to enter the slag, and the grain-refining action might be lost. Other ladle additions that are frequently used are boron to improve hardenability of steel and rare earths to improve the impact strength of cast steel. The mechanisms that lead to these improvements have not been fully explained.

Recently, the treatment of cast iron with magnesium or cerium to produce a ductile cast iron has assumed considerable importance. The addition of 0.02 to 0.10% magnesium or 0.02 to 0.04% cerium to a low-sulfur iron results in the formation of spheroidal rather than flake graphite in the as-cast material. The engineering properties of this alloy approach those of cast steel in many respects, and it is being used extensively for automotive crankshafts. The mechanism which operates in this case may involve changes in the surface tension; these may be brought about either by the direct action of magnesium or, indirectly, by a lowering of the oxygen level, which in turn may be due to the addition of magnesium. These changes result in a spheroidal, rather than a flake-type, mode of crystallization.

REFERENCES

1. D. P. SMITH, *Thermodynamics of Gas Metal Behavior.* A.S.M., Cleveland, Ohio, 1953, pp. 1–22.

2. L. W. EASTWOOD, *Gases in Non-Ferrous Metals and Alloys.* A.S.M., Cleveland, Ohio, 1953, p. 28.

3. D. J. CARNEY, J. CHIPMAN, and N. J. GRANT, "The Sampling and Analysis of Liquid Steel for Hydrogen," *Trans. A.I.M.E.*, **188**, 104 (1950).

4. *Basic Open Hearth Steelmaking.* A.I.M.E., 1952.

5. C. E. SIMS, H. A. SALLER, and F. W. BOULGER, "Effects of Various Deoxidizers on the Structure of Sulfide Inclusions," *Trans. A.F.S.* **57**, 233–248 (1949).

6. CYRIL STANLEY SMITH, "Grains, Phases and Interfaces, An Interpretation of Microstructure," *Metals Technology*, T.P. No. 2387 (June, 1948).

7. W. R. OPIE and N. J. GRANT, "Hydrogen Solubility in Aluminum and Some Aluminum Alloys," *J. Metals*, **188**, 10, 1237–1241 (1950).

8. D. J. CARNEY, J. CHIPMAN, and N. J. GRANT, "An Introduction to Gases in Steel," Electrical Furnace Steel Proceedings, *Trans. A.I.M.E.*, **6**, 34 (1948).

9. M. BEVER and C. F. FLOE, *Trans. A.I.M.E.*, **166**, 128 (1946).

10. A. J. PHILLIPS, "The Separation of Gases from Molten Metals," *Trans. A.I.M.E.*, **171**, 17–46 (1947).

11. C. W. BRIGGS, *Metallurgy of Steel Castings.* New York: McGraw-Hill, 1946.

12. H. EPSTEIN, J. CHIPMAN, and N. J. GRANT, "Hydrogen in Steelmaking Practice," *Trans. A.I.M.E.*, **209**, 597–608 (1957).

13. F. C. LANGENBERG, "Predicting the Solubility of Nitrogen in Molten Steels," *Trans. A.I.M.E.*, **206**, 1099–1101 (1956).

14. V. C. KASHIGAP, and N. PARLEE, "Solubility of Nitrogen in Liquid Iron and Iron Alloys," *Trans. A.I.M.E.*, **212**, 86–91 (1958).

15. W. S. PELLINI, W. H. JOHNSON, and H. F. BISHOP, "Improvement of Pressure Tightness and Tensile Properties of Gun Metal Bronze by Vacuum Degassing," *Trans. A.F.S.*, **63**, 345 (1955).

GENERAL REFERENCES

Basic Open-Hearth Steelmaking. A.I.M.E., 1951.

L. S. DARKEN and R. W. GURRY, *Physical Chemistry of Metals.* New York: McGraw-Hill, 1953.

O. KUBASCHEWSKI and E. L. EVANS, *Metallurgical Thermochemistry.* London: Pergamon, 1958.

R. SCHUHMANN, *Metallurgical Engineering.* Reading, Mass.: Addison-Wesley, 1952.

PROBLEMS

1. To degas liquid aluminum at 700°C (1292°F), impure argon containing 10% nitrogen and 2% hydrogen by volume is used. The degassing is continued to equilibrium with this gas at 1 atm. The degassing is then stopped, and the pressure over the melt is decreased continuously and rapidly with a vacuum pump. What are the changes that you will notice as the appearance of the melt varies as a function of the decreasing pressure? (Be quantitative in your answer.)

2. At 700°C (1292°F), a melt of pure aluminum contains 0.2 ml of hydrogen per 100 gm of metal. A bell jar is placed over the melt and evacuated at constant temperature (700°C). At a given pressure the melt apparently boils.

(a) Why does it boil?

(b) What is the pressure? (Neglect hydrogen diffusion to the atmosphere during pumping down.)

(c) To what value would you reduce the pressure in order to avoid porosity in a permanent mold casting poured at this temperature. (Neglect hydrogen segregation during freezing.)

CHAPTER 11

CONTROL OF CHEMICAL COMPOSITION

11–1 General. One of the most important applications of metallurgical thermodynamics is to indicate the factors which enable the engineer to keep the chemical composition within the desired limits. In this chapter we review the principles underlying the control of the common elements in ferrous and nonferrous metallurgy. Segregation during normal freezing and zone refining will also be discussed, in that order. The next chapter will deal with various types of furnaces and their design and operation, based upon the principles developed here. The reader is again reminded that a review of the calculations can be found in Chapter 13.

Chapter 10, on gases, has already laid the foundation for our present discussion. While the hydrogen, nitrogen, and oxygen contents of metals are rarely specified, we now know that they can drastically affect casting properties. As we increase our knowledge of the quantitative effects of these elements, we shall free ourselves more and more from the habit of tediously and inexactly relating the properties of the metal to the process used and to the "heredity" of the alloy. We will take up ferrous alloys first and then discuss the nonferrous alloys.

Ferrous Alloys

Our discussion of the factors governing control of the common elements in ferrous alloys will follow the order in which they usually appear on analysis sheets: carbon, manganese, silicon, phosphorus, sulfur, and alloying elements.

11–2 Carbon. The most valuable quantitative data for the control of the carbon content in liquid steel and gray and ductile iron are given by the iron-carbon equilibrium diagram and by the carbon-oxygen equilibrium in the metal (Figs. 12–6 and 10–13). The important effect of the activity of the oxygen in various slags will also be reviewed.

The iron-carbon diagram is basic to the study of carbon control, since it indicates the maximum solubility in iron of carbon in the form of coke in the blast furnace or cupola and graphite or other carburizers in the arc furnace. Although the diagram shows that the solubility of carbon increases with temperature, the equilibrium data do not give any information on the effects of such other important factors as the solution rates of different cokes and the wetting of coke by various slags. The

TABLE 11–1

SUMMARY OF CARBON-OXYGEN RELATIONS IN LIQUID STEEL
(Calculated from incomplete experimental data [1])

Temperature, $^{\circ}$C $^{\circ}$F			1500 2732	1550 2822	1600 2912	1650 3002	1700 3092
$10^3 \cdot m = 10^3 \cdot a_C \cdot a_O/p_{CO}$			2.03	2.17	2.30	2.45	2.60
$p_{CO}^2/p_{CO_2} \cdot a_C$			341	428	530	650	800
$p_{CO_2}/p_{CO} \cdot a_O$			1.45	1.07	0.81	0.63	0.49
w/o C	a_C		Carbon dioxide in equilibrium gas at 1 atm, w/o				
0.01	0.01		19.1	16.4	14.0	11.9	10.1
0.02	0.02		11.5	9.5	8.0	6.7	5.6
0.05	0.05		6.4	4.3	3.5	2.9	2.4
0.10	0.101		2.8	2.2	1.8	1.5	1.2
0.20	0.206		1.4	1.1	0.9	0.7	0.6
w/o C	f_O	f_C	Equilibrium product, [w/o C] $\cdot$ [w/o O] $\times 10^3$ at 1 atm CO + CO$_2$				
0.01	1.00	1.00	1.64	1.81	1.98	2.16	2.34
0.02	0.99	1.00	1.82	1.97	2.12	2.29	2.46
0.05	0.97	1.00	1.96	2.13	2.29	2.45	2.64
0.10	0.94	1.01	2.08	2.23	2.38	2.54	2.71
0.20	0.88	1.03	2.20	2.36	2.50	2.67	2.84
0.50	0.74	1.10	2.49	2.65	2.81	3.05	3.18
1.00	0.55	1.23	2.96	3.20	3.40	3.63	3.84
2.00	0.30	1.70	3.95	4.25	4.50	4.80	5.10

interrelation of dissolved carbon and oxygen is also vital. Although the equilibrium data are usually concerned with an atmosphere of CO, they apply quite well to ordinary operating conditions, as shown by the graph of Fig. 10–13.

In steel production, the key to carbon control lies in the C-O relations, whereas in high-carbon irons (malleable, cast, or ductile irons) the melting temperature of the iron and the characteristics of the coke are the dominant factors. The C-O equilibrium at different temperatures can be calculated with the aid of Table 11–1 [1]. As mentioned in the section on dissolved oxygen, the product of w/o C times w/o O at a given temperature is not constant. It is therefore necessary to correct the weight percent of each element by means of an activity coefficient to obtain the activity or

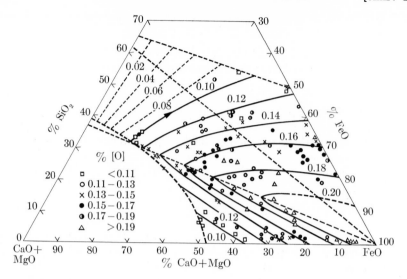

Fig. 11–1. Oxygen content of iron in equilibrium with synthetic slag at 1600°C (2912°F) [1].

"effective" concentration. In other words, because the reactive or escaping tendency is not equal to the weight percent, a correction factor is required for the calculation of each concentration.

If we set $p_{CO+CO_2} = 1$ atm and assume 0.5% carbon in steel at 1600°C (2912°F), we can calculate the equilibrium oxygen content of the melt as follows (see Table 11–1) for low weight percents:

$$a_C = \%C \cdot f_C = 0.50 \times 1.10 = 0.55,$$
$$a_O = \%O \cdot f_O = 0.74x,$$

where $x = \%O$ and f is the correction factor or activity coefficient. Then we have

$$10^3 \cdot a_C \cdot a_O / p_{CO} = 2.30$$
$$10^3 \times 0.55 \times 0.74x = 2.30,$$
$$x = 0.0056\%[O].†$$

(Study of the middle portion of Table 11–1 shows $p_{CO} \sim 1$ atm.)

The lowest portion of Fig. 11–1 illustrates a simplified calculation in the form of m-products, $\%C \times \%O = m$, which are not constants, i.e.,

---

† The reader is reminded that a symbol enclosed in brackets such as 0.1% [O] means 0.1% oxygen dissolved in the liquid metal. Where the symbol is in parentheses, such as 0.15% (FeO), it means dissolved in the slag. Mol fraction, instead of weight percent, is generally used in the case of slag.

they vary with % carbon. The above solution can be checked as follows:

$$(0.5 \text{ w/o C}) \times x[\%\text{O}] = 2.81 \times 10^{-3},$$

$$x = 0.0056\%[\text{O}].$$

These data also indicate that the reduction of carbon to fairly low levels, e.g., 0.10% C, as in plain carbon-steel practice, requires a moderate quantity of dissolved oxygen, i.e., 0.02%. Reduction of carbon beyond this point requires a very large quantity of dissolved oxygen, for example, that developed by either a high FeO slag or an oxygen lance.

The effect of slag on oxygen potential and hence upon carbon reduction is illustrated in Fig. 11–1. To reduce carbon to 0.02%, 0.10% [O], the quantity of dissolved oxygen required at 1600°C (2912°F) (Table 11–1), can be obtained from an FeO content of 15% in a basic slag of 35% SiO_2, 50% CaO + MgO. In an acid open hearth, considerably more FeO is needed, as shown by the slope of the 0.10% [O] curve in the same figure.

To recarburize to an accurate final carbon content, the dissolved oxygen must be taken into account, and if the oxidizing slag is left on the metal (single-slag practice), the oxygen in the slag must also be reckoned with. For this reason, the heat is often blocked (oxidation stopped) by a silicon or an aluminum addition which reduces the oxygen content of both bath and slag to a very low value. The carbon from the pig iron or wash metal which is added after the block is then completely recovered (see the deoxidation calculations in Chapter 10).

11–3 Manganese. Manganese is such a common element in steel and iron that most of our thinking is conditioned to an iron plus 0.50% manganese base rather than to pure iron. In malleable iron, the Mn-S balance affects the type of graphite formed; in spherulitic cast iron, Mn has a pronounced effect upon carbide stability, and in steel it is a useful ingredient in complex deoxidizers. It is worth while, then, to assemble here the available control data.

The vapor pressure of pure manganese is given by the Clapeyron equation [1],

$$\log p = \frac{-\Delta H}{4.575T} + B.^*$$

For manganese, $-\Delta H = 55{,}150$ cal/mole, $B = 4.97$, and therefore at 1600°C (2912°F), the vapor pressure is appreciable (0.035 atm). In a 1%-solution of Mn in liquid iron, if we consider the pressure to be

---

* With the values of B and ΔH given and with T in degrees kelvin, the pressure is in atmospheres.

proportional to the concentration, the vapor pressure of the manganese would be 0.00035 atm. Thus, when melting occurs in air at atmospheric pressure, the loss of Mn is small. In vacuum melting, however, the distillation of manganese is appreciable, particularly when the pressure is reduced below the vapor pressure of manganese.

In the usual furnace bath (open-hearth, air, or arc furnace) the following equilibrium calculation applies:

$$[Mn] + [O] = MnO \quad \text{(with FeO in slag)},$$

$$K_{Mn} = \frac{(MnO)}{\%[Mn] \times \%[O]}.$$

The MnO in the slag does not affect the *dissolved* oxygen, which depends upon the activity of the FeO in the slag. In slags containing only FeO and MnO the mol fraction of FeO in the slag may be substituted for the activity of FeO or for O. Thus we have

$$K = \frac{(MnO)}{(FeO)\%[Mn]},$$

which, if we use the data of Chipman [1], results in the empirical equation

$$\log K_{Mn} = +\frac{6640}{T} - 2.95,$$

or, at 1600°C (1873°K):

$$\log K = \frac{6440}{1873} - 2.95$$

$$= 0.48;$$

$$K = 3.02.$$

For example, if at 1600°C,

$$\frac{(MnO)}{(FeO)} = \frac{1}{5}, \quad \text{then} \quad \%Mn = 0.066.$$

The ratio (MnO/FeO) in the deoxidation product may also be calculated from the above equation; for example, at 0.4% Mn, the ratio is 1.21 and the mol fraction of MnO is 0.55.

We may combine this equation with another expressing the relationship of dissolved [O] to mol fraction (FeO) as follows:

$$(FeO) = [O] + Fe\ (l), \quad \text{Reaction 1},$$

$$Fe\ (l) + (MnO) = (FeO) + [Mn], \quad \text{Reaction 2},$$

Adding the two reactions, we have

$$(MnO) = [Mn] + [O], \quad \text{Reaction 3,}$$

$$K_3 = \frac{\%[Mn] \times \%[O]}{(MnO)} ;$$

For reaction 1,

$$\log K_1 = -\frac{6320}{T} + 2.734,$$

For reaction 2,

$$\log K_2 = -\frac{6440}{T} + 2.95.$$

Adding the two equations, we have

$$\log K_3 = -\frac{12760}{T} + 5.68.$$

The oxygen content in equilibrium with 0.4% Mn is calculated in the following way. Taking 1600°C (2912°F) as the basis, where (MnO) = 0.55, we find from the FeO-MnO-Mn relation that

$$K = 0.0742 = \frac{0.4 \times \%[O]}{0.55}, \quad \%[O] = 0.11.$$

This oxygen level can also be obtained in iron-carbon alloys with only 0.02% C. Manganese is therefore a rather weak deoxidizer and remains at moderate levels during the boil.

11–4 Silicon. Very rewarding information can be obtained from a study of silicon which lends itself readily to investigation. By calculations we will show, for example, how silicon can block a heat and protect carbon from oxidation, and how the silicon may be either reduced or oxidized, depending upon slag conditions. However, we do not yet know all there is to know about the effects of silicon. For example, a great many anomalies seem to be caused by the timing of the silicon additions, rather than by the final silicon content in the heat, and these are still to be explained. Although two cast irons may have the same final silicon content, a variation in graphitizing behavior leading to pronounced differences in hardness and strength can be demonstrated if the additions were made at different times. The later the addition, the lower the amount of iron carbide and the greater the amount of graphite in the final casting.

Let us first, as in the case of manganese, consider the equilibrium with dissolved oxygen and a silica slag:

$$(SiO_2) = [Si] + 2[O], \quad K = \frac{[Si] \times [O]^2}{(SiO_2)} .$$

We may calculate the equilibrium constant from thermodynamic data as follows. At $1600°C$ $(2912°F, 1873°K)$,

$$(SiO_2) = [Si] + 2[O],$$

$$\Delta F° = +129,440 - 48.44T.$$

For pure silica slag, the activity of SiO_2 is unity, and

$$K^* = 3.0 \times 10^{-5} = \%[Si] \times \%[O]^2.$$

To attain $0.10\%\,[O]$ during a boil, the $\%\,[Si]$ must be lowered to 3×10^{-3}, or 0.003%.

It is often considered that almost any set of conditions will permit oxidation of silicon. This is not true. Temperature, for example, has an important effect on the silicon-oxygen equilibrium. At $1700°C$ $(3092°F)$, the amount of silicon in equilibrium with oxygen in liquid steel is about six times greater than at $1600°C$ $(2912°F)$, and under mildly reducing conditions $(0.02\%\,[O])$, $0.45\%\,[Si]$ is the equilibrium value. In acid electric-furnace practice, at high temperatures, some reduction of silicon is obtained from the silica of the slag. However, when basic slags are used, this is uncommon, because both the actual mol percentage of SiO_2 and its activity are lower. The SiO_2 tends to bond with the basic constituents, and there is less (free) SiO_2 available for reaction. The equilibrium constant may still be used in the calculation, but a low value of a_{SiO2} must be substituted for the concentration of SiO_2. An important corollary of this effect is that a given silicon addition will provide more effective deoxidation under basic than under acid slags.

11-5 Phosphorus. Our chief interest in phosphorus is focussed on methods of eliminating it. The general equations may be written as follows:

$$4\,CaO_{slag} + 2[P] + 5\,FeO_{slag} = (CaO)_4P_2O_{5slag} + 5[Fe],$$

$$K' = \frac{(a_{(CaO)_4P_2O_5})(a_{Fe})^5}{(a_{[P]})^2(a_{FeO})^5(a_{CaO})^4}.$$

Since $a_{Fe} \cong 1$ and a_{FeO} is proportional to $\%\,[O]$ in low-carbon steel, we may simplify the above expression by gathering constants into K,

$$K = \frac{a_{(CaO)_4P_2O_5}}{(a_{[P]})^2(a_{[O]})^5(a_{CaO})^4},$$

---

* See Chapter 13 for details of calculation of K from ΔF.

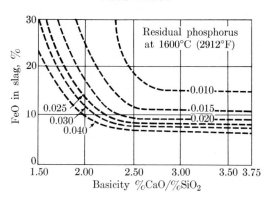

Fɪɢ. 11–2. Schematic curves showing approximate effect of slag lime/silica ratio and slag FeO on residual phosphorus for conventional open-hearth practice [1].

or, alternatively, we may write:

$$\frac{a_{\text{phosphorus in slag}}}{(a_{\text{phosphorus in metal}})^2} = K(a_{[\text{O}]})^5(a_{\text{CaO}'})^4.$$

The CaO is now written CaO′ to denote free or unbonded lime which is available for reaction (to be discussed later).

For good dephosphorization, i.e., for a high distribution ratio of $P_{\text{slag}}/P_{\text{metal}}$, it should be noted that the influence of oxygen increases with the fifth power of the oxygen concentration and that of free lime with the fourth power of lime concentration.

The difficult part of this problem, quantitatively, is to calculate the activity of CaO. Just as in the case of SiO_2, bonding with other slag constituents lowers the reactivity of CaO. This problem has been solved [2] with a series of assumptions concerning slag composition which are found to be justified by a plot of the resulting data according to Raoult's law, and we obtain the following equilibrium constant equation:

$$\log K = \frac{71{,}667}{T} - 28.73.$$

However, this formula is very inconvenient to use since the CaO′ of a given slag must be calculated by a tedious trial-and-error solution of a fourth-degree equation. Therefore, prepared graphs of both temperature and lime effects are employed to estimate dephosphorization. Figure 11–2, for example, shows that there is little benefit in exceeding a basicity ratio (% CaO/% SiO_2) of 2.5, but that dephosphorization continues to increase with increasing FeO for an FeO-content above 25%. With these data, the slag weight for a given dephosphorization ratio may then be calculated.

11–6 Sulfur. It has been demonstrated [3] that sulfur is removed from liquid metal by two mechanisms.

(1) The slag dissolves iron sulfide from the metal until equilibrium is reached, i.e., until the tendency of FeS to escape from the metal equals the tendency of FeS in the slag to escape to the metal. In other words,

$$FeS_{metal} = FeS_{slag}.$$

(2) Iron sulfide reacts with lime (CaO) at the slag-metal interface to produce calcium sulfide and iron oxide:

$$FeS_{metal} + CaO_{slag} = CaS_{slag} + FeO_{slag}.$$

If the iron-oxide level of the slag is already high, very little CaS can be found in the slag, and reaction (1) predominates.

Since the desired end result is the accumulation of sulfur in the slag and since it does not matter whether the sulfur appears there as FeS or CaS, we may, after analyzing a given metal-slag combination for sulfur, express the desulfurizing power as %S slag/%S metal. This ratio may vary from as high as 400:1 to as low as 1:1,* depending on the conditions of the given situation.

It is important to realize that this ratio is not constant for a given slag if the base composition of the metal is changed. In other words, the activity of the FeS and hence its tendency to leave the metal vary with the composition. For example, when carbon and silicon contents are high, as in cast iron, FeS fortunately has a greater tendency to leave than when these elements are present in small amounts [4, 5]. For a given slag, the ratio (%S slag/%S metal) is four times higher for a 3.6% C, 1.5% Si cast iron than for pure iron. To compare data obtained from metals of differing compositions "%S" must be converted to activity a_S [5]. The ratio (%S slag/a_S metal) obtained for different metal compositions with the same slag is then constant, and the action of the slag alone may be studied.

The two principal factors affecting slag desulfurization are iron oxide content and basicity, as shown in Fig. 11–3 [3]. From the chemical analysis, iron oxide is computed as %FeO + 0.9% Fe_2O_3. Percent Fe_2O_3 is adjusted by 0.9 to equalize the molar %O in Fe_2O_3 with that in FeO. Free base is computed by the following steps.

---

* The use of a ratio of this sort is simpler in evaluating slags since the relative weights of slag and metal are eliminated from consideration and may be substituted in later calculations. For example, assume that the average sulfur content of a scrap charge is 0.10% and that a final content of 0.02% is desired in the metal. If equal weights of slag and metal are employed, a desulfurization ratio of 0.08/0.02 = 4 will be satisfactory if equilibrium is attained. If, for practical operating conditions, a ratio of only 5 lb of slag to 100 lb of metal is to be used, a desulfurization ratio of 80:1 will be required under equilibrium conditions.

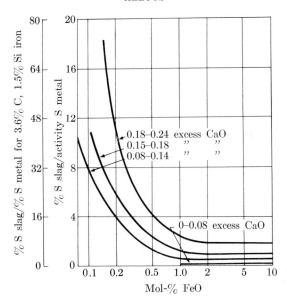

FIG. 11–3. Effect of slag analysis on desulfurization [1].

(1) First the basicity ranking b is determined:

$$b = \frac{CaO + MgO}{SiO_2 + Al_2O_3} \quad \text{(use molar percentages)}.$$

(2) When b is less than 2.3, bases (CaO and MgO) are assumed to combine with acids (SiO_2 and Al_2O_3) on a 1 : 1 molar basis. After proportionate amounts of CaO and MgO have been used to neutralize the acids, the remaining CaO is considered to be free lime.

(3) When b is greater than 2.3, bases (CaO and MgO) are assumed to combine with acids (SiO_2 and Al_2O_3) on a 2 : 1 molar basis, and the residual CaO is again considered to be free lime.

In the basic cupola, both low iron oxide (below 2%) and high basicity may be obtained, while in the acid cupola lower basicity and sometimes higher oxidation are encountered. The effect of iron oxide is illustrated by Fig. 11–3, and the effect of varying %CaO versus %SiO_2 in the slag is demonstrated by the excess base calculation.

11–7 Alloys. Before discussing the problems involved with individual alloying elements, we may profitably review the situation from an overall point of view. Let us consider the various ways of arriving at the alloy content of a widely used low-alloy steel such as AISI 4340 (0.40%C, 0.7% Mn, 0.04% maxP, 0.04% maxS, 0.3%Si, 1.8%Ni, 0.8%Cu, 0.25% Mo). The simplest yet most expensive method would be to melt a commercially pure grade of iron and add relatively pure carbon, manganese, silicon, nickel, chromium, and molybdenum. However, the premiums paid for the elements in the pure form would be excessive, and it is far more practical to add their ferro-alloys which are cheaper and dissolve more readily than the pure metals.

Assuming that we have poured castings of this composition, the question arises as to the disposition of the gates and risers. Can we salvage the alloy content by remelting? It is obvious that this problem becomes even more acute in making highly alloyed castings. In addition, scrap materials, such as the various grades of stainless steel, can often be obtained at prices below the value of the alloy in the scrap. What will happen to these elements during the CO boil or when dephosphorizing and desulfurizing slags are used?

The answers to these questions can be found in the thermodynamics of the reactions between the alloying elements and other elements such as oxygen. We may divide these effects into three groups:

1. The element remains substantially in the metal bath regardless of slag or oxidation (Ni, Cu, Mo).

2. The element enters the slag in large quantities under oxidizing conditions but remains in the bath under a reducing slag (Mn, Si, Cr).

3. The element is very readily oxidized (V, Al, Ti, Cb, W) and, once in the slag, is not reduced to the bath.

Group 1. Ni, Cu, Mo. This group is very simple to deal with. If these elements are present in the scrap, they will be found in the finished metal, or in other words, the distribution ratio of the element between slag and metal is very small. Furthermore, if an oxide such as nickel oxide or a complex salt such as calcium molybdate is added to the bath, the metal (Ni or Mo) will be reduced. As an example let us calculate the equilibrium for nickel.

The reaction [Ni] + [O] = (NiO) (in slag) proceeds far to the left even if the oxygen level in the metal is high. We may obtain a quantitative relation of interest from the data given in Table 13–5. By adding and subtracting equations (all at 1600°C, 2912°F), we have:

$$\text{Ni (l)} = \text{[Ni]} \qquad \Delta F° = -17{,}300 \text{ cal/mole,}$$

$$+ \tfrac{1}{2}O_2 \text{ (g)} = \text{[O]} \qquad \Delta F° = -28{,}990 \text{ cal/mole,}$$

$$\overline{\text{Ni (l)} + \tfrac{1}{2}O_2 \text{ (g)} = \text{[Ni]} + \text{[O]}} \qquad \Delta F° = -46{,}290 \text{ cal/mole,}$$

$$(-)\ \text{Ni (l)} + \tfrac{1}{2}O_2 \text{ (g)} = \text{(NiO)} \qquad \Delta F° = -13{,}650 \text{ cal/mole,}$$

$$\overline{\text{(NiO)} = \text{[Ni]} + \text{[O]}} \qquad \Delta F° = -32{,}640 \text{ cal/mole;}$$

$$\log K = \frac{32{,}640}{(4.575)(1873)} = 3.81, \quad K = 6.47 \times 10^3,$$

whence we obtain the expression:

$$\frac{a_{[\text{Ni}]} \times a_{[\text{O}]}}{a_{\text{NiO}}} = 6470.$$

In other words, if we assume as a first approximation that the activities are proportional to the concentrations, then the amount of nickel in the liquid will be far greater than the NiO content in the slag—even at high oxygen levels in the metal. Hence, if an alloy heat is lanced with gaseous oxygen and the slag is then removed, little nickel will be lost. Similar reasoning applies to copper and molybdenum.

Group 2. Mn, Cr, Si. These elements are readily oxidized, in fact, it can be calculated that a major "fuel" for the converter is silicon, as well as carbon. The loss of manganese and silicon, relatively inexpensive elements, is generally tolerated during the oxidizing period since it is considered more important to provide a good CO boil for flushing hydrogen and nitrogen and oxidizing phosphorus. Since Mn and Si reactions have already been discussed, we need not dwell on these points.

The oxidation of chromium, however, is extremely important. With the advent of 12–14% Cr and 18 Cr–8 Ni stainless steels, large quantities of stainless steel scrap have become available. At the same time it is found that the corrosion resistance is inversely proportional to the carbon content. Hence, specifications with carbon below 0.07% are now common. The melting problem thus becomes a question of how to reduce the carbon without oxidizing the chromium during refining. When ordinary techniques such as the use of iron ore are tried, a thick refractory slag rich in chromium oxides is formed before the carbon is reduced. However, through the injection of oxygen, the resulting high bath and slag temperatures permit carbon oxidation with lower losses of chromium to the slag. Furthermore, the chromium can be reduced from this hot fluid slag later in the heat by adding silicon. The interrelation of carbon and chromium contents in the metal as a function of temperature is shown in Fig. 11–4 [7]. The figure is based on the general equation

$$\log \frac{\%[\text{Cr}]}{\%[\text{C}]} = \frac{-13,800}{T} + 8.76,$$

from which it follows that the higher the temperature, the smaller the negative term on the right-hand side of the equation and the higher the Cr/C ratio. Of course, the maximum temperature is limited by the furnace-roof life, and an interesting economic balance develops between the cost of scrap versus virgin materials and operating costs.

Group 3. V, Al, Ti, Cb, W. With the exception of aluminum all the alloys listed are relatively expensive. When small amounts not exceeding

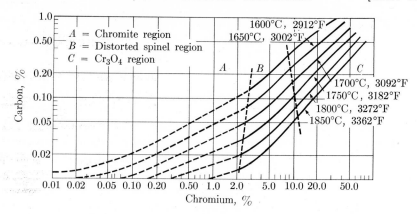

FIG. 11–4. Carbon-chromium relations as affected by temperature [7].

two percent are involved, the melt is usually made in the same way as an unalloyed heat, and the ferroalloys of the above elements are added after deoxidization with silicon or aluminum. Sometimes a reducing slag, high in carbon or calcium carbide and very low in FeO, is melted first to provide a protective cover. On the other hand, when large quantities of these elements are contained in the scrap, as is the case with highly alloyed tool steels containing over 5% W and 2% V, a simple reducing slag is used for dead melting. However, sometimes a brief oxygen injection is applied to reduce the hydrogen content. Fortunately in tool steels and in highly alloyed cast heat-resistant alloys, a higher carbon content (over 0.4%) is desired.

Another method of degassification, namely melting or casting under vacuum, leads to exceptionally good recovery of the oxidizable elements except those with high vapor pressure.

NONFERROUS ALLOYS

11–8 Nickel-base alloys. We may profitably discuss the melting practice for the nickel-base alloys next because of the similarity with the techniques used for iron and steel alloys. The essential difference between nickel and iron-base alloy melting is that the usual starting material, electro-nickel squares or carbonyl nickel, is far purer than that used for commercial iron-base alloys. The carbon-oxygen reaction is similar to that in liquid steel, and additions of carbon and nickel oxide are made deliberately to produce the boil required for hydrogen reduction. Sulfur content is carefully controlled and kept to a maximum of 0.01%, by proper selection of raw materials and avoidance of high-sulfur fuels when a reverberatory furnace is used for melting. Lead is also a deleterious contaminant and should be kept below 0.01%. A thin lime slag is generally employed as a cover.

After the heat has been degassed by the carbon boil, the oxygen is reduced by the addition of about 0.3% metallic silicon, and the slag is kept in a reducing condition by means of small calcium-silicon additions. The carbon level is raised with low-sulfur charcoal, and then silicon, manganese, and other elements are added to reach the desired composition. The heat is finally deoxidized with metallic magnesium.

Deoxidation by means of magnesium is in interesting contrast to steelmaking practice. It is introduced for two reasons: (1) Magnesium is far more soluble in liquid nickel than in liquid steel. (2) The tapping temperature of nickel alloys is over 100°F (56°C) lower than that of steel so that the vapor pressure of magnesium is smaller. The analysis control for high chromium-nickel alloys such as Inconel (18% Cr, balance Ni) and high copper alloys such as Monel (30% Cu, balance Ni) is similar, but the deoxidation with magnesium is usually accomplished in the furnace.

11–9 Copper-base alloys. For copper-base alloys, composition control is ordinarily less of a problem than for the materials previously discussed because ingot of the desired final analysis is generally available. The usual batch weight of copper-base alloys is small since crucible melting is extensively employed. Hence one might expect to encounter many cases in which it would be necessary to weigh out small amounts of alloying elements and to incur a high cost of analysis for small heats. Instead, the secondary copper smelter oxidizes and refines large quantities of scrap in a reverberatory furnace, makes the required alloy additions, and produces ingot of the desired composition for remelting. The relative merits of oxygen versus air in the elimination of impurities such as Al, Si, Mn, P, and Fe are illustrated in Fig. 11–5 [8].

According to the types of problems presented by the control of both the base composition and the gas content, copper-base alloys may be divided into three groups: (1) Copper castings containing only small amounts of alloying elements (used when high thermal and electrical conductivities are required); (2) copper-tin-lead-zinc alloys and copper-zinc (up to 10% Sn, 10% Pb, 40% Zn); (3) copper-aluminum, copper-silicon, copper-beryllium, and copper-nickel alloys.

(1) The key to compositional control for copper castings often lies in the use of minimum amounts of alloying elements and deoxidizers to provide maximum conductivity. The principles have been discussed in Chapter 10. The technique consists of the following steps: (a) melting clean scrap and virgin copper in an oxidizing atmosphere free from SO_2, (b) deoxidation with phosphorus, leaving a minimum residual amount, and (c) using dry-sand molds to minimize hydrogen pick-up. Lithium

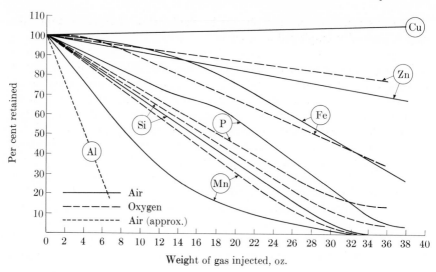

FIG. 11–5. Relative oxidation of foreign elements in copper ingots. Starting percentages: Cu 81.4, Sn 3.4, Pb 3.4, Ni 0.76, S 0.12, Sb 0.12, Al 0.06, Fe 1.23, Si 0.335, Mn 0.12, P 0.12, Zn 8.9 [8].

or calcium boride may be substituted for part or all of the phosphorus when maximum conductivity is desired.

(2) Copper-tin-lead-zinc alloys have a lower hydrogen solubility than copper (see Chapter 10). The precautions to be observed in melting these alloys resemble those for copper except that a small excess of phosphorus is allowable since conductivity is not, as a rule, important. The oxygen content never reaches the high values obtained with pure copper because of the buffering effect of the other elements. The only element usually lost during melting is zinc. At the 5%-level, as in the common 85 Cu, 5 Sn, 5 Pb, 5 Zn alloy, the oxidation loss is about 0.5% Zn. Just before deoxidization this loss is made up. Then 0.02% P in the form of a 15% P-Cu alloy is added. It is desirable to practice careful scrap segregation and to reserve certain crucibles exclusively for these base alloys since substantial amounts of Si, Al, Fe are reported to cause difficulty.

When the lead content is over 10%, it is important to superheat the alloy to over 2200°F (1204°C), to ensure that all the lead is dissolved and that none is present as a second liquid phase.

(3) The composition control of copper-aluminum, copper-silicon, copper-beryllium and copper-nickel alloys presents difficulties because of density differences in the alloying elements. Uniform solution is accomplished by use of ingots of the desired composition or by the addition of master alloys. Careful scrap segregation is again necessary because the

presence of even small amounts of other elements can result in "hard spots" due to foreign phases (usually high in iron). The copper-nickel alloys are relatively simple to control because of the ready solubility and similar density of the elements; however higher melting temperatures are required.

11–10 Aluminum-base alloys. The chief problem in the compositional control of these alloys is hydrogen, and this aspect has been discussed at length in Chapter 10. Contamination by iron in certain high-strength alloys such as A356 is to be avoided. The aluminum alloys are generally made from custom-smelted ingot although in some cases the melter will make selected additions. If copper or silicon is added, precautions should be taken to ensure thorough solution. Manganese, nickel, titanium, and chromium should be added in the form of master alloys. When magnesium is to be added, it should be plunged to minimize oxidation loss.

11–11 Magnesium-base alloys. As in the case of aluminum, magnesium alloys dissolve hydrogen (Chapter 10). Degassing is accomplished with chlorine since nitrogen reacts with magnesium. Because of the low density, oxides and other nonmetallic phases may be trapped in the melt. The use of a magnesium-potassium chloride flux in conjunction with agitation by chlorine helps to clean the metal. The only impurity that is picked up in conventional operations is silicon which results from the reduction of silica sand. Adjustment of composition is accomplished by master alloys, except for aluminum and zinc additions which are made in the elemental form.

11–12 Zinc-base alloys. Most specifications require the use of virgin materials for zinc die castings, and this requirement is justified by the deleterious effects of small amounts of Pb, Sn and Cd on intercrystalline corrosion. Most zinc alloys are melted in iron pots, and overheating must be avoided to minimize solution of iron.

ZONE REFINING

An increasing number of applications reveal that the properties of a metal are not linearly related to impurity content but may vary in a more complicated way. Examples are the conductivity of metals at low temperatures, the fascinating properties of the semiconductors in transistors, the effects of small amounts of dissolved gases on the properties of titanium and chromium. Since the time may be near when considerable quantities of castings of exceptionally high purity may be needed, it seems worth while to discuss a new technique which has been developed to reduce impurity levels—to as low as 0.000001% in some cases. This technique, called zone refining, may be applied to castings of simple shapes by means of a refractory permanent mold. A related procedure which avoids segregation of alloying elements is called zone leveling.

Before discussing these processes in detail, we need to review the distribution of an element B dissolved in a liquid A as solidification takes place. Assume that we pour a simple square bar (Fig. 11–6a) and chill one end so that it freezes with a plane front proceeding from the chilled to the hot end. We neglect freezing from the insulated sidewalls. Let

C_0 = original concentration of B in liquid A,

C_L = concentration of B in liquid at any time t ($C_L = C_0$ at $t = 0$)

C_s = concentration of B in solid at the solid-liquid interface,

x = distance of solid-liquid interface from chilled end,

L = total length of bar,

g = fraction frozen = $\dfrac{x}{L}$,

a = cross-sectional area,

$k_0 = \dfrac{Cs}{C_L}$ under equilibrium freezing conditions,

$k = \dfrac{Cs}{C_L}$ under nonequilibrium freezing conditions.

Both k_0 and k are taken to be constant for the freezing range under consideration. This assumption is not valid for all systems, but is useful for small concentrations of B. We also assume for present purposes that the diffusion is very fast in the liquid compared to the solid. Therefore, the liquid is of uniform composition at any time t, whereas the solid composition varies with x but does not change at any point x after freezing. We can now derive the relation [9]

$$\frac{C_s}{C_0} = k_0(l - g)^{k_0-1}.$$

If C_s is plotted as a function of g (Fig. 11–7) for different values of k_0 or k [10], it is evident that the solidified material near the chilled end contains much less of the solute (impurity) B. The values are not plotted beyond $g = 0.9$ because the equation does not hold for final solidification. The important point, however, is that starting with $C_0 = 1\%$, the im-

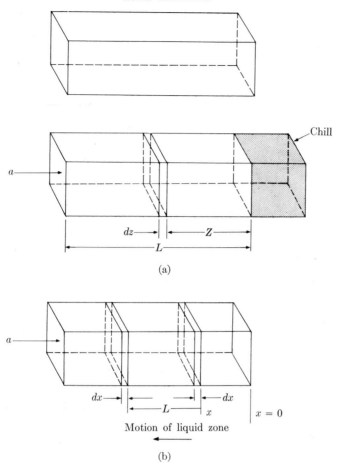

FIG. 11-6. (a) "Normal" segregation. (b) Zone refining.

purity level is reduced to below 0.1% for 0.9 of the bar length when k_0 or $k = 0.01$. Even when this value is 0.5, considerable purification can be accomplished by directional freezing. It follows, then, that if we wished to purify a metal, we could cut off the end containing the high amount of B, then remelt and resolidify several times, always discarding the material which solidifies last.

Zone refining is essentially a means of accomplishing this purification without discarding the last fraction to solidify each time the bar is melted. However, to prevent the bar from being recontaminated, only a "zone" of length l is melted at a time, starting at one end and progressing to the other. The principle is illustrated in Fig. 11-6(b). The original bar can be prepared by any conventional means such as casting, rolling,

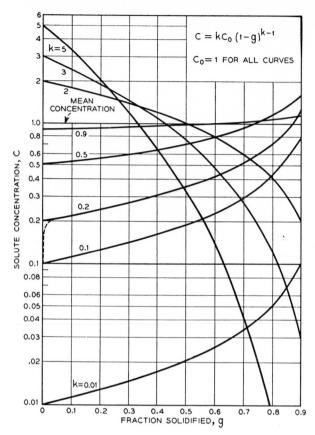

FIG. 11–7. Solute distribution with directional solidification [10].

etc. It is then supported in a tube or boat and passed through an induction, arc, or resistance heater which melts a section of length l. As the bar moves through the hot zone, the section leaving the zone freezes. The solidified section does not have the same content of B as the liquid in the molten zone because the composition of the solid is a function of k as well as of C_L. When k is less than unity, the freezing solid contains less B than the liquid does. Consequently, the liquid becomes richer in the impurity and, in effect, sweeps it to the end of the bar. Using our previous terminology, we see that if we start at one end of the bar and melt a zone of length l, the concentration in the liquid will be initially C_0. As the liquid zone moves on, the first solid to precipitate will be of composition kC, and the equivalent volume of solid which melts at the other end of the zone is of composition C_0. Inasmuch as the volumes are equal, the liquid must become enriched in B.

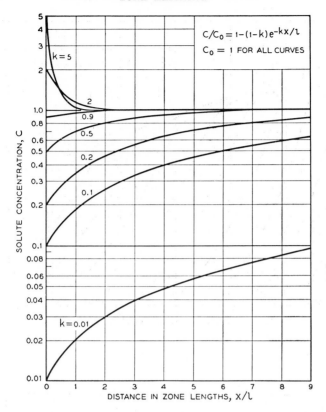

$$\frac{C}{C_0} = 1 - (1-k)e^{-kx/l}$$

$C_0 = 1$ FOR ALL CURVES

FIG. 11–8. Solute distribution with zone refining [10].

The equation for the composition of the solid at any distance x after the first pass is

$$\frac{C_s}{C_0} = 1 - (1 - k)e^{-kx/l}.$$

The difference between this equation and that for natural freezing arises from the effect of the molten zone of length l. The change in solute concentration is plotted as a function of distance in zone lengths in Fig. 11–8.

The effect of repeating the cycle is shown in Fig. 11–9 for (a) $k = 0.5$ and (b) $k = 0.1$, after different numbers of passes [10]. Note that with $k = 0.1$ and after 10 passes, the impurity level at the "pure" end of the bar has fallen to 10^{-7} times the initial impurity content.

In actual practice, the number of passes required to reach this low impurity level is considerably larger than in our example, which is based on the assumption that the concentration of B in the liquid is uniform.

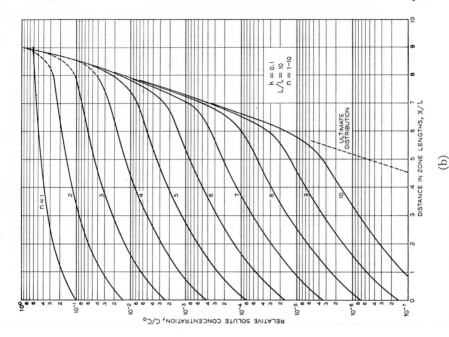

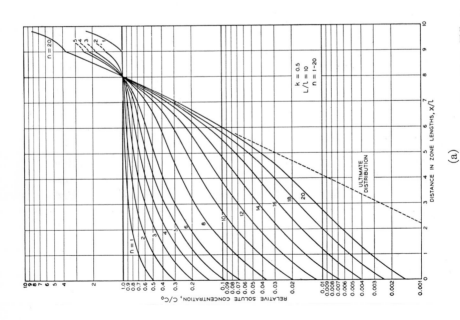

FIG. 11-9. Effects of repeated zone refining [10].

Actually B piles up at the solidifying interface (see discussion of constitutional supercooling in Chapter 2). It is possible to determine an "effective" k_e representing the actual ratio of C_s/C_L at the interface by the methods described in reference 12. However, the pile-up can be reduced by inductive stirring or by a supplementary magnetic field.

REFERENCES

1. *Basic Open Hearth Steelmaking.* A.I.M.E., 1952.

2. T. B. WINKLER and J. CHIPMAN, "An Equilibrium Study of the Distribution of Phosphorus Between Liquid Iron and Basic Slags," *Trans. A.I.M.E.*, **167**, 111–133 (1946).

3. R. ROCCA, N. J. GRANT, and J. CHIPMAN, "Distribution of Sulfur Between Liquid Iron and Slags of Low Iron Oxide Concentrations," Symposium on Desulfurization, A.I.M.E. (February, 1951).

4. G. G. HATCH and J. CHIPMAN, "Sulfur Equilibria Between Iron Blast Furnace Slags and Metal," *Trans. A.I.M.E.*, **185**, 274–284 (1949).

5. J. P. MORRIS and R. C. BUEHL, "The Effect of Carbon Upon the Activity of Sulfur in Liquid Iron," *Trans. A.I.M.E.*, **188**, 317 (February, 1950).

6. EISAMIN, "Electric Furnace Steel," *Proc. A.I.M.E.*, **7**, 252 (1946).

7. D. C. HILTY, H. P. ROSSBACH, and W. CRAFTS, "Observations of Stainless Steel Melting Practice," *J. Iron and Steel Inst.* 177 (June, 1955).

8. M. GLASSENBERG, L. F. MONDOLFO, and A. H. HASSE, "Refining Secondary Copper Alloys," *Trans. A.F.S.*, 465–471 (1951).

9. W. G. PFANN, *Trans. A.I.M.E.*, **194**, 747 (1952).

10. W. G. PFANN, *Zone Melting.* New York: Wiley, 1958.

11. W. A. TILLER, "Solute Segregation During Ingot Solidification," Scientific Paper 8–0108–P11, Westinghouse Research Laboratory.

GENERAL REFERENCES

Transactions A.I.M.E., Transactions Brit. Iron and Steel Inst., and references listed for Chapter 10.

PROBLEMS

1. Assume that you wish to make ductile cast iron, using a basic-lined direct-arc furnace. The final composition of the iron to be tapped from the furnace is 3.6% C, 1.6% Si, 0.02% S. The most convenient economical source of raw materials is a machine shop whose borings average 3.7% C, 1.7% Si, 0.12% S. From previous experience you have developed a slag containing 0.2 mol-percent FeO and 0.21 "excess CaO."

What weight of slag per ton of metal would be required?

2. Calculate the w/o FeO needed to produce 0.1% [O] in a slag containing 50% SiO_2, balance FeO and (CaO + MgO), at 1600°C (2912°F).

3. Check the statement that at 1700°C (3092°F) 0.45% Si is the equilibrium value with 0.02% [O] in a silica slag.

CHAPTER 12

SELECTION AND CONTROL OF MELTING PROCESSES

12–1 General. The development engineer is often faced with the problem of deciding which furnace or combination of furnaces should be used for a new alloy or for a standard material in a new installation. The production staff should also be alert to the effects that the type of furnace and lining may have upon production costs. For example, let us suppose that some defect such as a hot tear accounts for a 5% scrap loss, and that it can be shown that a revised composition, with lower sulfur and phosphorus content, would reduce this loss substantially. Even though the additional refining required may add to the melting expense, new lining and slagging techniques that result in considerable scrap reduction could lead to a worth-while over-all saving.

In this chapter we discuss the proper selection of a furnace for a particular process and present a tabular comparison of the types currently available (Table 12–1).

Although furnaces differ greatly in design, kinds of fuel required, and cost, there are really only a few key factors which, in addition to production requirements and casting specifications, govern the choice of furnace type. These factors are:

(a) *Metal chemistry.* This includes not only the control of the standard elements but also that of gases and certain "hereditary" effects, such as the relative machinability of Bessemer and open-hearth metal, and the difference in the chilling tendencies exhibited by cupola and electric-furnace cast iron respectively. The size, shape, and composition of the available raw materials are also important considerations.

(b) *Metal temperature.* For delicate castings or complicated metal transfer systems, the metal in the ladle must be maintained at very high temperatures. When a decision has to be made with respect to the optimum melting equipment for a new material, the fluidity relationships developed in Chapter 5 are often quite useful in estimating the required temperature. The maximum tapping temperature consistently obtainable from a given furnace may be a governing factor in the selection of the appropriate furnace type.

(c) *Metal delivery rate.* Two aspects are of vital importance to foundry planning: the rate of metal delivery and the mode of delivery (whether the molding calls for batch or continuous melting). When large castings, say rolls, are made, the batch process is preferable, since the time required to accumulate sufficient metal from a continuous unit such

TABLE 12–1

FURNACES FOR FERROUS MELTING

Furnace	Metal temperature		Metal delivery rate, tons/hr or tons/batch	Operating cost, three shifts†	Capitalization‡ at high and low delivery rates
	°F	°C			
Cupola	2850–3000	*	Cont. 50–2	$8–$15/ton	$50,000–$4,000
	2850–3000	1566–1649	Cont. 50–2	$9–$13	$50,000–$4,000
Air furnace	2850–3000		Batch 50–10	$12–$14	$100,000–$20,000
Open hearth	2900–3000	1593–1649	Batch 50–10	$16–$20	$200,000–$100,000
	2900–3000		Batch 50–10	$18–$22	$220,000–$100,000
Direct arc	2850–3200	1566–1760	Batch 25–1/2	$17–$25	$250,000–$50,000
	2850–3200		Batch 25–1/2	$19–$27	$250,000–$50,000
Indirect arc	2850–3100	1566–1704	Batch 2–0.1	$20–$40	$20,000–$8,000
Induction (H.F.)	2850–3200	1566–1760	Batch 5–0.1	$30–$45	$70,000–$25,000
	2850–3200		Batch 5–0.1	$32–$47	$70,000–$25,000
Converter	3000–3100	1649–1704	Batch 5–1	$16–$18	$10,000–$5,000

Table 12–1 (*cont.*)

Metal-chemistry control ranges

Furnace	Lining	% C	% Mn	% P§	% S§	% Si
Cupola	Acid	3.8–2.5	4.0 max.	0.05 min	0.08 min	4.0 max
	Basic	4.4–2.7	4.0 max.	0.04 min	0.01 min	4.0 max
Air furnace	Acid	3.6–1.5	optional	0.04 min	0.03 min	optional
Open hearth	Acid	1.5–0.05	optional	0.03 min	0.03 min	optional
	Basic	1.5–0.02	optional	0.02 min	0.03 min	optional
Direct arc	Acid	3.8–0.02	optional	0.02 min	0.03 min	optional
	Basic	3.8–0.02	optional	0.01 min	0.01 min	optional
Indirect arc	Acid	3.8–0.50	optional	0.03 min	0.03 min	optional
Induction (H.F.)	Acid	4.2–0.02	optional	0.03 min	0.03 min	optional
	Basic	4.2–0.02	optional	0.02 min	0.02 min	optional
Converter	Acid	0.20–0.02	0.1 max	0.04 min	0.04 min	0.1 max

* With preheated air, calcium carbide, or oxygen.
† Includes depreciation expense. These values are not exact and may vary considerably with fuel and labor costs. Where costs are of the same order local conditions may therefore reverse the position of several of these types of furnaces.
‡ Furnace alone; no charging or auxiliary equipment, since this cost is exceedingly variable.
§ Note especially that these values are not obtained by refining high-sulfur and phosphorus scrap as can be done with basic refractories.

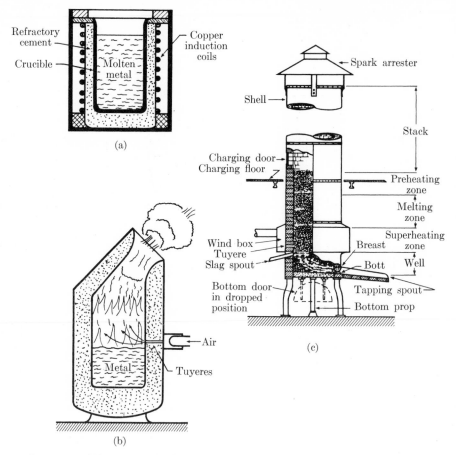

Fig. 12–1. Diagrammatic sketch of principal melting furnaces. (a) Induction furnace. (b) Side-blow converter. (c) Cupola.

as a cupola might seriously affect the pouring temperature. On the other hand, for a conveyorized operation for small castings a large heated holding furnace is needed in conjunction with batch melting, whereas a smaller, insulated ladle usually suffices for continuous melting.

(d) *Melting cost.* The balance between capital investment and operating cost is often critical. Shortly before the depression of 1932, a number of foundries scrapped cupolas, bought electric furnaces, and based their calculations of depreciation on the full capacity of the plant. They came up with a figure which amounted to only a nominal cost per ton until operation was cut back to one or two days a week; then depreciation became a prohibitively high portion of the cost. In a new plant the melting equipment is a major item of expense, and hence, unless existing facilities are being used to capacity 24 hours per day, further capital expenditures

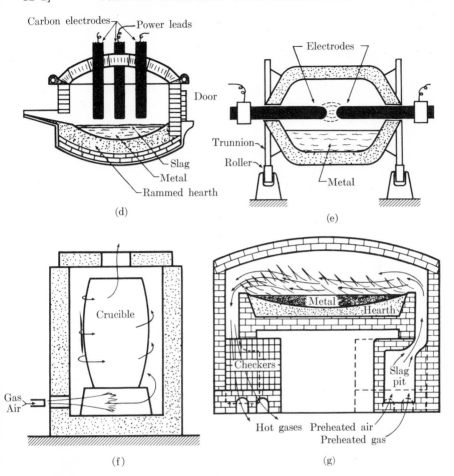

FIG. 12–1 (*cont.*). (d) Direct-arc furnace. (e) Indirect-arc furnace. (f) Gas-fired crucible. (g) Open-hearth furnace.

should not be considered. On the other hand, with assured high-tonnage production, low operating costs can permit rapid write-off of very sizable capital investments.

We first compare the operating characteristics of various furnaces, and then discuss the chemical control of the processes involved.

12–2 Overall comparison of melting furnaces. At first glance, the schematic drawings of various types of melting furnaces shown in Fig. 12–1 seem to be quite complex and appear to have little in common. However, they can be subdivided into two simple principal groups: furnaces for batch melting and for continuous melting.

(a) *Batch processes.* In virtually all batch processes a bath is heated by oil, gas, or electricity.* The oldest type of furnace is the crucible furnace, and crucible melting is the most common method for batch melting of nonferrous alloys. The crucible is heated by the flame and radiation from the sidewalls, and transfers the heat to the bath. This method, however, is no longer used for steel; induction furnaces are the modern substitute. In ferrous induction melting the heat is developed directly in the metal, instead of being applied externally to a container. In nonferrous induction melting, a conducting crucible of graphite or silicon carbide is used because the resistance of the melt itself is too low for proper induction heating.

Reverberatory, open-hearth, and air furnaces in which the flame heats the bath both directly and by radiation from the roof are other long-established types. In these furnaces, considerable length is required to attain proper heat transfer from the long flame. The direct-arc furnaces, on the other hand, do not require the length of the open hearth, since the electrodes deliver the electrical energy locally. In indirect-arc furnaces, the heat of the arc radiates to the sidewalls, which are rocked under the metal. If cold metal is placed between the electrodes of an indirect-arc furnace, the first stage of melting is due to electrical resistance.

(b) *Continuous melting processes.* The most widely used continuous melting unit is the cupola, which has no real bath. As shown in Fig. 12–1, the metal charge is melted by the hot gases from the coke bed, and the metal droplets receive additional heat as they fall through the bed.

The coke consumed during melting is replaced by adding fresh coke between metal charges. The slag which forms above the metal in the well of the cupola is derived from flux, such as limestone, which is added with the coke. The slag can also trap material such as coke ash and remove sulfur from the iron droplets as they pass through. The cupola is mainly used for melting cast iron, but it is also employed for brass and bronze. For iron the usual melting rate is 20 pounds per minute per square foot of melting area, where the melting area is taken to be the area of the cross section above the tuyeres.

12–3 Chemical control in cupola melting. Until about 1955 the cupola was considered merely as a cheap melting instrument, and cupola slag was treated as a necessary evil to be kept at a minimum. With the advent of spherulitic graphite irons, however, controlling the sulfur and carbon

---

* Except for the special case of a converter, which is really a reaction vessel and not a melting unit. In this process, the temperature is increased by combustion of Si, C, and Fe, and other elements in the steel.

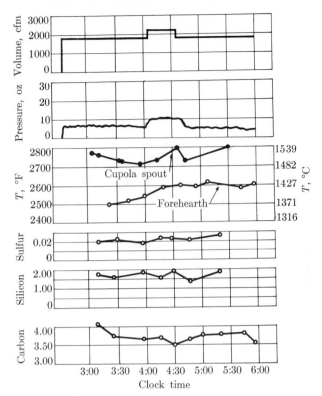

Fig. 12–2. Log sheet of basic cupola heat [4].

content of the iron became economically imperative. Following early work in England, a number of investigators in this country made a concerted attempt to develop an economical basic cupola operation [1 through 5].

In the basic cupola the fireclay refractories are replaced by either magnesite brick, monolithic magnesite ramming mix, or by a water-cooled shell which develops a thin slag layer (basic) from the limestone and other fluxes. The monolithic magnesite is superior to the brick which spalls readily. A cold blast can be used with the monolithic magnesite, but a hot blast is usually required with the water-cooled shell.

Control of sulfur and phosphorus. When magnesium or cerium is added to gray iron, a desulphurizing reaction occurs first and only then can magnesium or cerium be retained in the iron, in the amount required to control graphite shape. An excess of either element leads to high chill and is also an added expense. Since approximately one part of magnesium combines with one part of sulfur and since a Mg residual of about

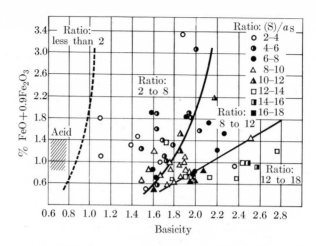

FIG. 12–3. Effect of slag composition on desulfurization under cupola conditions [4]. Basicity = $(CaO + \frac{2}{3}MgO)$ $(SiO_2 + Al_2O_3)$, molar percents.

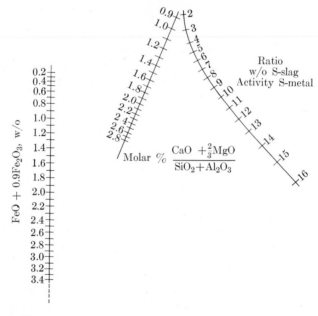

FIG. 12–4. Nomograph for calculating the effect of slag composition on desulfurization [4].

0.06% is desired, it is obvious that using a 0.06% S base iron, instead of 0.01% S iron, will lead to operating difficulties.

As shown in Fig. 12–2, basic cupolas can be operated for sulfur contents ranging from 0.01 to 0.02% by means of slag control. As discussed in Section 11–6, control depends upon the amount of FeO and the basicity of the slag. Accordingly, various different slags have been experimentally melted in basic cupola operations to determine the resulting ratio (% S slag/% S metal). Inasmuch as carbon and silicon affect the activity of the sulfur in the slag by a factor of 4 in this composition range, a given desulfurization ratio for steel, % S slag/% S metal, will be increased four times for cast iron. It is easier, therefore, to obtain a low sulfur content in a high-carbon iron. In analyzing the results, one plots sulfur-distribution ratios versus FeO and basicity, as shown in Fig. 12–3. The plot provides a good correlation even though equilibrium conditions in the cupola cannot be guaranteed. The reason for the poor desulfurization of acid slags is also apparent from the plot and the nomograph of Fig. 12–4.

Phosphorus behavior is plotted in Fig. 12–5. While the slags are sufficiently basic for dephosphorization, reference to Fig. 12–5 shows that the accompanying FeO and resulting [O] are inadequate for appreciable P-reduction. By the use of lower operating temperatures and an oxidizing blast, it has been possible to accomplish appreciable dephosphorization [1]. The difficulty encountered in attempts to remove sulfur and phosphorus simultaneously arises because of the conflicting effects of dissolved oxygen.

Carbon control. The carbon contents obtainable with different types of cupola operations are sketched in Figs. 12–6 and 12–7. The problem of carbon control is quite different from that of sulfur and phosphorus control, and the liquid iron is usually quite far from equilibrium. The solution of carbon from coke by molten iron depends on the composition and temperature of the iron, the coke-iron interface conditions, and the type of coke used. In general, both hot-blast and basic cupolas have higher melting-zone temperatures than the simple acid-lined cupolas, and carbon solution is therefore increased. In addition, the adherence of the slag to the coke is reduced in the basic cupola and a more reactive coke surface is available.

Summary. The acid cupola is cheaper and easier to operate when no refining is needed and some sulfur increase (0.02 to 0.10%) can be tolerated. The basic cupola provides low-sulfur iron, can carburize the liquid iron to higher levels, and shows some promise of dephosphorization *under certain operating conditions.* However, dephosphorization cannot occur under desulfurizing conditions.

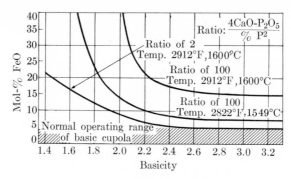

FIG. 12–5. Effect of slag composition on dephosphorization [5]. Basicity = $(CaO + \frac{2}{3}MgO)(SiO_2 + Al_2O_3)$, molar percents.

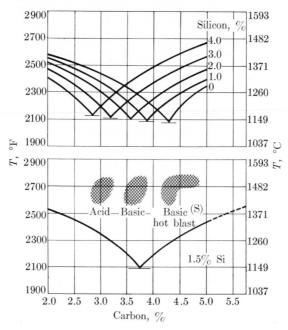

FIG. 12–6. Effect of silicon and operating conditions on carbon solubility [4].

The following general formula for acid melting has been developed [2]:

$$\text{Total C of melted metal} = 2.4 + \tfrac{1}{2}C - \tfrac{1}{4}(Si + P),$$

where the quantities of Si, C, and P given apply to an ingoing charge in the following composition range:

%Si	%P	%C
0.9 to 2.5	0. to 0.65	1.1 to 3.9

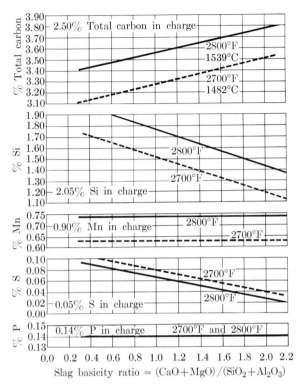

FIG. 12–7. Effects of slag basicity and tapping temperature on composition of iron for cupola charges containing 35% steel (excludes carbon linings and calcium carbide).

12–4 Chemical control in the air furnace and the open hearth.

The chief difference between air furnaces and the open hearth is the higher operating temperature of the latter, produced by preheating the air in a regenerative checkerwork. Air furnaces are used principally for the melting of malleable iron, and to some extent for gray iron, while open-hearth operations are confined to steel castings. These acid and basic open-hearth operations produce about 500,000 tons per year.

An air furnace is essentially a melting or holding furnace, and, except for a small carbon decrease (0.10 to 0.50% C) at the level of 2.0 to 3.0% carbon, no important changes in standard analysis take place. The chill of the metal does increase markedly in the case of malleable iron, and this effect may be related to solution of hydrogen from furnace gases or to the slagging off of nuclei for graphitization.

The acid open hearth may be used as a cold-melting furnace or in conjunction with hot metal from the cupola. A review of the require-

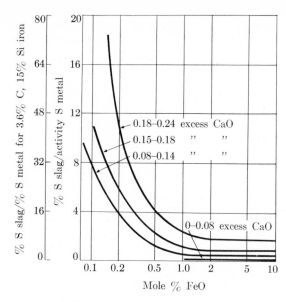

FIG. 12–8. Effect of slag composition on desulfurization [8].

ments for sulfur and phosphorus reduction (Chapter 11) indicates that little can be accomplished with typical acid finishing slags having the following composition:

%SiO$_2$	%FeO	%MnO	%CaO	%Al$_2$O$_3$
50	20	10	8	7

The open-hearth process consists primarily of charging an iron-steel mixture averaging 1.0% carbon, using the carbon boil to ensure low hydrogen and nitrogen content and to provide good heat transfer, and finally deoxidizing with silicon, silicomanganese, aluminum, or titanium. The fact that about 1.0% carbon is necessary in the charge to produce 0.30% carbon steel is economically important. Because steel scrap has an average carbon content of about 0.30%, either relatively high-grade pig iron or cupola metal is required for the balance, since sulfur and phosphorus cannot be eliminated. In times of pig-iron shortage, this situation has imposed a serious burden upon the acid open-hearth operator.

The acid heat provides an interesting application of the FeO-MnO-[Mn] equilibrium discussed in Chapter 11. Recall that at low carbon levels the oxygen content of the metal under an FeO-MnO slag is proportional only to the FeO content; this is in line with the observation that high MnO slags permit closer control of the carbon drop in the working and finishing periods of the heat. With a high FeO/MnO ratio, oxidation (an endothermic reaction) can occur at too rapid a rate, resulting in a cold bath and a gummy slag.

Basic open-hearth chemistry is exhaustively described in *Basic Open Hearth Steelmaking* [6]. Certain special conditions develop in the manufacture of castings. For example, while the basic open hearth is an excellent dephosphorizing medium, it also desulfurizes, but, as shown by Fig. 12–8, the FeO content is too high to develop a really satisfactory desulfurization ratio. With a ratio (% S/% [S])* of 4, an absurd slag weight of 25% of the metal weight would be necessary to reduce the sulfur level of the bath from 0.06 to 0.03%. The finishing of a basic heat requires great care to prevent phosphorus reversion to the bath. For instance, an excess of silicon can lower the quantity of dissolved oxygen and reduce the $(CaO)_4P_2O_5$ in the slag. Some plants use only manganese in the furnace and add silicon to the ladle.

In general, open-hearth operations are useful for low-cost batch melting of steel in large quantities. Acid-furnace operation is slightly cheaper if pig iron and scrap of low phosphorus and sulfur contents are available.

12–5 Chemical control in arc melting. Acid electric practice has an advantage over the open-hearth method in that the development of a gummy slag can be prevented by using the arc to provide the high temperatures required to keep the slag fluid. As a result, a much lower carbon content may be charged, and the cycle is shortened. The acid elec-

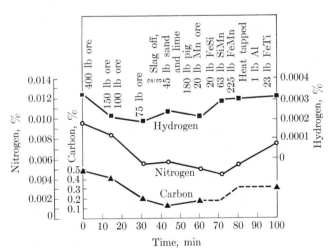

Fig. 12–9. The change in carbon, nitrogen, and hydrogen contents during the production of a six-ton acid-arc furnace heat [10].

---

* Brackets indicate that the element is dissolved in liquid metal.

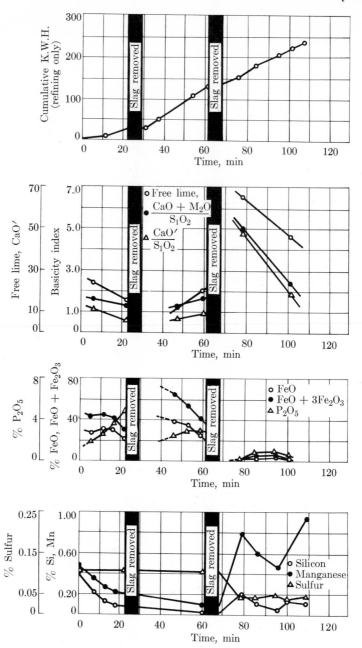

Fig. 12–10 (a). Phosphorus, sulfur, and carbon reactions in basic electric-arc furnace [10]. (See Fig. 12–9 for operating data of this heat.) (*cont.*)

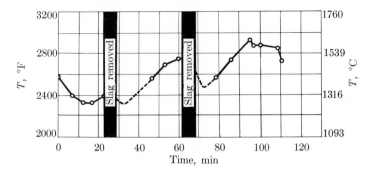

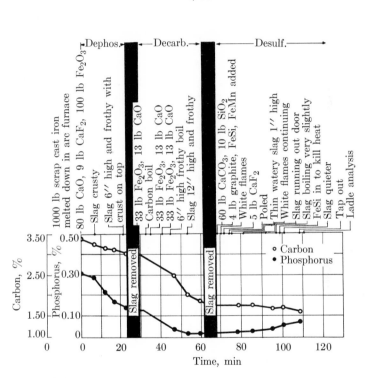

FIGURE 12–10 (b) (cont.)

tric practice is essentially a fast-melting, nonrefining method. The melt may either be decarburized beyond the point aimed for and carburizer may be added, or it can be blocked with silicon at the proper carbon level (established with the guidance of a rapid carbon-analysis technique). Silicon reduction can be accomplished (Chapter 11) by the reaction $2\,[\mathrm{C}] + \mathrm{SiO_2} = [\mathrm{Si}] + 2\,\mathrm{CO}$, but there is a difference of opinion as to the advisability of this practice. Although a reduced slag provides better recovery of deoxidizers, the conclusion of the boil renders the heat susceptible to gas pickup. Bath temperatures of 3100 to 3200°F (1704 to 1760°C) are needed for reduction. The variation in hydrogen, nitrogen, and carbon during a typical acid heat is illustrated in Figs. 12–9 and 12–10(a).

The basic electric-arc furnace is perhaps the most versatile melting tool at the disposal of the foundryman. In double-slag practice, both phosphorus and sulfur may be reduced to very low levels, and with the help of such other devices as the oxygen lance, remarkable use of high-alloy scrap may be made. In double-slag practice, a basic, high FeO slag is employed to reduce carbon and phosphorus to the desired values. The FeO can be provided either by ore additions, by melting with an open door, or by using oxygen. The oxygen lance results in an exothermic reaction and short heats, while Fe_2O_3 additions are endothermic and lead to longer melting times. This first slag is then pulled, and a reducing basic slag is substituted to accomplish desulfurization. The great difference in desulfurization between basic open-hearth and electric-furnace slags is due to the low FeO content of the latter.

The graph of Fig. 12–10(b) is an interesting illustration of these effects. The data shown were obtained from a special-purpose heat that was run to determine the possibility of producing a 1.5% C steel, low in phosphorus and sulfur, from cast-iron scrap with 0.30% P and 0.12% S. It is evident that by using low temperatures, the phosphorus may be reduced to low levels before decarburization, as indicated by the difference in slope of the equilibrium constant curves. Dephosphorization decreases with increasing temperature, whereas the carbon reaction requires high temperatures. The pronounced drop in silicon and manganese indicates that these elements are removed before the dissolved oxygen in the metal can reach a high enough value to permit efficient phosphorus oxidation. The removal of sulfur under the second slag with reducing conditions is also noteworthy.

Indirect-arc melting. The indirect-arc furnace (also called the rocking furnace) is essentially a low-cost batch furnace for small melts. The smallest *direct*-arc furnace which can be conveniently operated with automatic controls has a melting capacity of about 500 pounds. Indirect-arc furnaces, on the other hand, can be easily operated with charges as small as 200 pounds, and so are employed for a variety of special alloys. The maximum permissible temperature in an indirect-arc furnace is lower than that for a direct-arc operation because there is a tendency to burn out the roof. The slower melting rate, the tendency of slag to build up on the walls, and the lower temperature range restrict the use of indirect-arc furnaces to melting and permit little refining. However, these furnaces are well suited to the lower-melting copper-base alloys and alloyed cast irons.

12–6 Chemical control in induction furnaces. The induction furnace is an ideal melting tool when it is required that the charge have approximately the same composition as the casting. Aside from gas elimination, very little refining is ever attempted, because the slag is colder than the metal and the protective refractory of the furnace walls is thin. Slag erosion, accelerated by the stirring action of the induction current, can cause early failure of the furnace walls. Furnaces of this type are usually small, but they are important when protective atmospheres or vacuum melting is desired. Low-frequency (60 to 180 cycles) induction furnaces are receiving increasing attention, particularly as holding furnaces for cupola metal.

12–7 The converter. The converter provides a startling example of the reaction rates which are possible at steelmaking temperatures if the reactants are well mixed. In comparison to the four-hour cycle in the open hearth or the one-hour cycle in the arc furnace, the converter reactions are completed in a matter of minutes. The graph of Fig. 12–11 shows the rapid oxidation of C, Si, and Mn and the solution of nitrogen

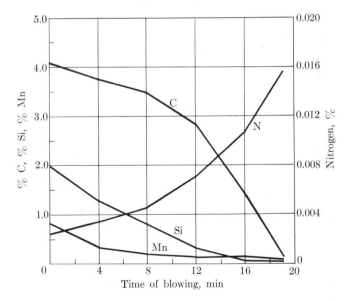

Fig. 12–11. Composition change during Bessemer blow; silicon substantially oxidized at 1.5% carbon [10].

during a characteristic blow in an acid converter. Here again no refining is accomplished because of the acid lining.

Basic converters are more widely used in Europe than in this country since their heat is derived from the oxidation of 1.5 to 2% phosphorus, and ores of this type are not available here. Satisfactory operation has been obtained by blowing oxygen-enriched air from a water-cooled tuyere above the bath in the BOF converter [9]. The principal factors limiting converter use have been smoke, high metal loss (10 to 15%), and nitrogen content, but the use of oxygen reduces the latter two markedly.

12–8 Duplex and triplex processes. Instead of regarding the various furnaces as isolated melting instruments, the foundryman should consider them as a tool kit designed to assist him in obtaining the best operating balance among chemistry, temperature, delivery rate, and cost. Each situation merits careful consideration, not only to determine the proper process for optimum production, but also because plant location may depend upon such factors as freight or power rates.

A great many successful installations have used the cupola as the primary melting instrument, even in steelmaking, because of the low level of both operating cost and capitalization. There is little doubt that the direct-arc furnace is a very flexible means of attaining the desired temperature and composition, but its output can be doubled or trebled if hot metal is provided from another source. In refining operations, it is often better to limit the refining to one furnace and use a second furnace for holding and superheating.

12–9 Nonferrous melting furnaces. The principal conditions governing the selection of a nonferrous melting furnace are usually different from those in ferrous practice, since the chemistry of nonferrous metals is controlled by the selection of proper melting materials, and only rarely is refining a decisive factor. Once a source of satisfactory ingot or scrap has been found, the principal problems derive from the presence of gases in the melt.

For the light metals, aluminum and magnesium, the refining process is usually limited to the use of gaseous or solid fluxes to trap dross and heavy metal impurities and to remove gas. Crucible melting is the commonest method. (As previously discussed, control of furnace gases is desirable because they can be a source of hydrogen, water vapor, or sulfur dioxide.) Small reverberatory furnaces are also used, however, for greater economy.

For copper-base alloys, oxidation is occasionally conducted in a reverberatory furnace to arrive at a desired composition; the melting cost is low for high-volume production. High-frequency induction furnaces have an excellent record for the melting of copper of high conductivity. The indirect-arc furnace and the crucible are the principal methods in the field of brass and bronze melting.

REFERENCES

1. S. F. CARTER, "Basic Lined Cupola for Iron Melting," *Trans. A.F.S.* **58**, 376–502 (1950).
2. W. W. LEVI, "Melting Iron in a Basic Lined Water Cooled Cupola," *Trans. A.F.S.* **60**, 740–753 (1952).
3. E. S. RENSHAW, "Basic Cupola Melting and Its Possibilities," *Trans. A.F.S.* **59**, 20–27 (1951).
4. R. A. FLINN and R. W. KRAFT, "Importance of Slag Control in Basic Cupola Operation," *Trans. A.F.S.* **59**, 323–331 (1951).
5. R. DOAT and M. A. DEBOCK, "Metallurgical Blast Cupola," *Trans. A.F.S.* **60**, 44–52 (1952).
6. *Basic Open Hearth Steelmaking.* 2nd ed. A.I.M.E., 1951.
7. L. S. DARKEN and R. W. GURRY, *Physical Chemistry of Metals.* New York: McGraw-Hill, 1953.
8. R. ROCCA, N. J. GRANT, and J. CHIPMAN, "Distribution of Sulfur Between Liquid Iron and Slags of Low Iron Oxide Concentrations," Symposium on Desulfurization, A.I.M.E. (February, 1951).
9. C. D. KING, "Steelmaking Processes—Some Future Prospects," *Trans. A.I.M.E.* (Metals Branch) **200**, (1954).
10. T. SWINDEN and F. B. CAWLEY, *Iron and Steel Ind.* **12**, 387 (1939).

GENERAL REFERENCES

Cupola Handbook, A.F.S., 1952.
Transactions of Electric Furnace Conferences, A.I.M.E.
Copper Base Alloy Foundry Practice, A.F.S., 1952.
See also references in Chapter 11.

PROBLEMS

1. Select melting equipment and materials for the following products and state the principal reasons for your choice and any necessary qualification (such as slag, specific raw materials, etc.).

Casting	Quantity	Composition
Automobile engine blocks	5000	C 3.25%, Si 2.0%, Mn 0.5%, S 0.15% max, P 0.15% max
Liners for gyratory crushers (weight 400 lb)	100	C 0.5%, Si 0.5%, Mn 1.2%, P 0.05% max, S 0.05% max

TABLE 12–2. MELTING STOCK

	C	Mn	S	P	Si	Cr	Ni	Mo	Approximate cost, $/ton
Armco iron	0.02	0.01	0.019	0.007	0.015				100
#1 pig	4.49	0.11	0.017	0.026	0.70				70
#2 pig	4.56	0.19	0.01	0.033	0.42				72
#3 pig	4.23	0.82	0.033	0.16	1.74				58
85% Ferrosilicon					85.18 bal Fe				0.19/lb*
50% Ferrosilicon					50.0				0.15/lb
Ferromanganese	7.00	75.0			1.00				0.16/lb
Manganese metal	0.20%	95.5			1.00				0.46/lb
Ferrochrome (low carbon)	0.03				0.38	70.6			0.41/lb
Ferrochrome (high carbon)	6.00				8.50	64.0			0.22/lb
Nickel							98.0		0.81/lb
Ferromolybdenum					0.41			60.6	1.76/lb
Silicon metal					97.0				0.25/lb
SAE 1018 scrap	0.2	0.6	0.03	0.03	0.1				60
Ferrophosphorus				24.0					120
Aluminum ingot	99+ Al								0.27/lb
Iron sulfide									
Pure copper ingot									0.30/lb
Aluminum (wire)									0.30/lb
Calcium boride	30% Ca, 40% B, 17% C								
Ce-Mg-Fe-Si	45.99 Si, 9.18 Mg, 0.98 Al, 0.57 Ce, 1.00 Ca								0.25/lb
Nickel-magnesium	71.01 Ni, 19.86 Mg, 3.94 Cu								0.70/lb
Aluminum bronze ingot	85.88 Cu, 3.48 Fe, 10.38 Al								

* Per pound contained element in all cases.
Note: Ferrosilicon briquets contain 1 lb silicon. Ferromanganese briquets contain 2 lb manganese.

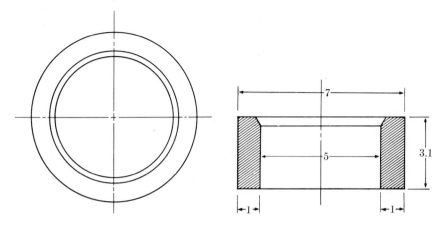

Fɪɢ. 12–12. Valve seat insert.

2. ꜰʀᴏᴍ: Chief Engineer.
   ᴛᴏ: Metallurgist.
   ꜱᴜʙᴊᴇᴄᴛ: New ring inserts for Supertronic Mfg. Co.
We have just been asked to bid on the rings shown in the attached sketch (Fig. 12–12). The composition is given below.

C	Mn	P	S	Si	Ni	Cr	Mo
2.8	0.7	0.15	0.15	2.0	4.0	5.0	1.0
		max	max				

In order that we may arrive at a reasonable quotation, I would like you to prepare a detailed specification for an initial order of 100,000 in a one-month period. Please cover the following points.

(a) Mold design for maximum yield. As a first approximation, use steel risering calculations.

(b) Specify melting practice, calculating charge and materials to be used. Use Table 12–2.

(c) Which of our complete selection of furnaces do you plan to use and with what melt weight, pouring temperature, etc.?

3. In the production of ductile iron, the sulfur content of the metal from the cupola should be 0.02% max. Assuming that the desulfurization ratio, % S slag / % S metal, is 40, and that the coke weight is equal to 15% of the metal charge, what is the final slag weight per 100 lb of metal required for the conditions given below.

S content of coke 0.5%, ash 8% $(Al_2O_3 + SiO_2)$; charge analysis:

C	Mn	P	S	Si
4.0	0.6	0.10	0.08	2.0

CHAPTER 13

REVIEW OF METALLURGICAL CALCULATIONS

13-1 General. In 1888, LeChatelier leveled the following criticism at the steel industry:

"It is known that in the blast furnace the reduction of iron oxide is produced by carbon monoxide according to the reaction

$$Fe_2O_3 + 3CO = 2Fe + 3CO_2,$$

but the gas leaving the stack contains a considerable proportion of carbon monoxide, which thus carries away an important quantity of unutilized heat. Because this incomplete reaction was thought to be due to an insufficiently prolonged contact between carbon monoxide and the iron ore, the dimensions of the furnaces have been increased. In England they have been made as high as thirty meters. But the proportion of carbon monoxide escaping has not diminished, thus demonstrating, by an experiment costing several hundred thousand francs, that the reduction of iron oxide by carbon monoxide is a limited reaction. Acquaintance with the laws of chemical equilibrium would have permitted the same conclusion to be reached more rapidly and far more economically."

Today, criticisms of the same sort can be leveled at the foundry industry for the many expensive, so-called "practical," tests of tuyere design, fluxes, ladle additions, and furnace practice which are contrary to the laws of metallurgical thermodynamics. Similar criticisms could be directed at investigations which are being conducted on a qualitative basis, where the operator is merely told to use an "oxidizing" or a "basic" slag, and so is led to awkward and wasteful operations and incomplete or incorrect conclusions.

The approach based on the laws of chemical equilibrium is so powerful because it enables the engineer to (1) know that a given reaction will or will not take place, (2) calculate the equilibrium conditions in advance, and (3) estimate the direction the investigation should take and the important variables to be studied even if experimental data are not available.

In this chapter we have consolidated some of the typical calculations that are used in the planning and analyzing of the processes of melting, refining and casting metals. No attempt has been made to derive the relations, however, since this is the task of a complete text on thermodynamics. Our discussion may be broken down into seven major parts:

(1) Review of units.

(2) The gas laws relating pressure, volume, temperature, and weight of gases.

(3) Mass and energy balances (first law of thermodynamics).

(4) The combined first and second laws of thermodynamics, and the concept of free energy as employed in calculating the possibility of various refining reactions, as well as the equilibrium constants and their variation with temperature.

(5) Gas-liquid equilibria that govern the relations among concentration of solutes, vapor pressures, temperature, and gas solubility.

(6) Slag-metal equilibria and partition of components, such as sulfur between slag and metal, and effects of slag and metal compositions upon thermodynamic activity.

(7) Combined use of phase diagrams and equilibrium constants. Concentrations of gas-forming impurities in the liquid phase during solidification, and buffering effects of deoxidants.

A word of caution should be added at this point. These relations tell us of the *possibility* of a reaction, not the *rate* at which it will occur. Fortunately, at liquid-metal temperatures, diffusion (a major rate-limiting factor) is rapid, and most reactions approach equilibrium rapidly.

13–2 Review of units. *Atomic and molecular weights.* From elementary chemistry, we recall that calcium oxide, CaO, does not contain equal *weights* of calcium and oxygen even though the ratio of calcium to oxygen *atoms* is 1 : 1. Instead, the ratio in which the two weights are combined is 40.08 to 16, which is the ratio of atomic weights. (Recall that O^{16} is the reference point for the other atomic weights of Table 8–1.) In the reaction of calcium with oxygen dissolved in liquid nickel, we have

$$Ca + [O]† = CaO.$$
$$\underset{\substack{\text{Dissolved in} \\ \text{liquid nickel}}}{} \quad \underset{\text{Slag}}{}$$

Since 40.08 grams of calcium can react with 16 grams of oxygen to form 56.08 grams of CaO, we see also, as a corollary, that CaO is

$$\frac{40.08}{56.08} \times 100 = 71.5\% \text{ calcium}, \qquad \frac{16}{56.08} \times 100 = 28.5\% \text{ oxygen},$$

and therefore CaO = 71.5% calcium + 28.5% oxygen.

Conversion of weight to volume. For problems in which we are given the weight of a solid or liquid which may be converted to a gas that is more easily measured as a volume, it will be recalled that one gram-mole

---

† Brackets indicate that the element is dissolved in liquid metal.

(molecular weight expressed in grams) of an ideal gas occupies 22.4 liters under standard conditions of temperature and pressure (0°C and 760 mm Hg). For example, since bromine vapor consists of Br_2 molecules, 79.916×2 grams of bromine will occupy 22.4 liters when the bromine is volatilized to approach ideal gas behavior, under standard conditions.

The basis for this general principle is, of course, that one gram-mole of any substance will contain the same number of molecules as one gram-mole of any other substance, i.e., 6.0235×10^{23}. It follows, then, that when any substance behaves like a perfect gas, a gram-mole contained in a volume of 22.4 liters at 0°C will have a pressure of 760 mm Hg (1 atm). For more precise conversions required by nonideal behavior, any standard physical chemistry text may be consulted.

13–3 The gas laws. The equation of state for an ideal gas is given by

$$pV = nRT,$$

where

$$p = \text{pressure, atm,}$$

$$V = \text{volume, cm}^3,$$

$$T = \text{absolute temperature, °K,}$$

$$n = \text{the number of moles of gas,}$$

$$R = \text{the universal gas constant.}$$

Recall that

$$\text{Kelvin temperature} = °C + 273 = \frac{°F - 32}{1.8} + 273,$$

and

$$R = 82.05 \frac{\text{cm}^3 \cdot \text{atm}}{\text{mole} \cdot °K} = 1.99 \frac{\text{cal}}{\text{mole} \cdot °K} \cdot$$

EXAMPLE. A tank containing 2.2 lb of dry nitrogen is to be used to purge a 220-lb melt of copper alloy, of density 8 gm/cm³. What is the volume of nitrogen at 2200°F (1204°C) compared with that of the melt when the nitrogen is bubbled through the melt at 2200°F?

Solution.

$$2.2 \text{ lb} = 1000 \text{ gm} \times (1 \text{ mole } N_2)/28 \text{ gm} = 35.7 \text{ moles } N_2,$$

$$\frac{(2200°F - 32)}{1.8} + 273 = 1477°K.$$

To determine the volume of the gas at 2200°F, we write

$$pV = nRT,$$

$$V = \frac{nRT}{p}$$

$$= \frac{35.7 \text{ moles}}{1 \text{ atm}} \times 82.05 \frac{\text{cm}^3 \cdot \text{atm}}{\text{mole} \cdot {}^\circ\text{K}} \times 1477°\text{K}$$

$$= 4{,}310{,}000 \text{ cm}^3 \quad \text{or} \quad 4.3 \times 10^6 \text{ cm}^3.$$

Hence the volume of the melt is

$$\frac{220 \times 453 \times 1}{8} = 12{,}400 \text{ cm}^3 \quad \text{or} \quad 1.24 \times 10^4 \text{ cm}^3,$$

and the volume of $N_2 = (4.3/1.24) \times 10^2 = 347$ times the volume of the melt.

PROBLEM. Calculate the relative volumes of the following gases compared with the metal, i.e., calculate the volume of 100 gm of copper and the volume of SO_2 formed by 0.01 gm of sulfur, etc.

0.0001% S by weight as SO_2 in copper at 2000°F (1093°C),
0.0001% H_2 by weight as H_2 in aluminum at 1400°F (760°C),
0.0001% H_2 by weight as H_2 in steel at 2700°F (1482°C).

Dalton's law of partial pressures is given by

$$p_t = p_1 + p_2 + p_3 + \cdots,$$

or

p_t (total pressure) = sum of partial pressures of all gases present, with each gas exerting the pressure it would develop if present alone.

EXAMPLE. Three gas cylinders of equal volumes contain hydrogen at a pressure of 50 psia (pounds per square inch absolute), CO at 20 psia, and nitrogen at 32 psia. The contents of the CO and nitrogen cylinders are pumped into the hydrogen cylinder to produce a mixed gas atmosphere for melting experiments. What is the pressure in the hydrogen cylinder?

Solution. The pressure will be the sum of the partial pressures of each gas acting independently, or 102 psia. (This assumes ideal gas behavior.)

PROBLEM. A crucible of liquid iron is in equilibrium with an atmosphere containing 20% N_2 and 10% H_2 by volume, the balance being made up by argon. Nitrogen and hydrogen can dissolve in the iron, the argon is insoluble. The melt is placed in a closed chamber which is evacuated rapidly at the original equilibrium temperature. What changes occur in the melt as the pressure drops?

13–4 Mass and energy balances (first law of thermodynamics). *Mass balances.* The practical operator might paraphrase the first law of thermodynamics as follows: "Everything that goes into the furnace must come out except that which stays in." This statement is illustrated by the following case. A particular foundry was using two cupolas of identical construction to melt cast iron containing about 0.15% sulfur. Curiously, however, for a period of several weeks one cupola produced metal about 0.04% higher in sulfur content than that produced by the other, although the metal charges, coke, air blast, and operating conditions were identical. A material balance of sulfur-in versus sulfur-out confirmed that somehow the first cupola was adding 0.04% sulfur, whereas in the second cupola, sulfur-in equaled sulfur-out. The only possible source of the additional sulfur was the refractory lining of the first cupola, which was indeed high in sulfur as shown by chemical analysis. A cost-minded purchasing agent had obtained a "bargain" in second-hand brick from an oil refinery working with high sulfur oils! Similarly, in melting 12% chromium heat-resistant steel, 0.5 to 2% nickel may be picked up from the lining of a furnace which has been used previously for 18% chromium-8% nickel alloy. Mass balance, therefore, serves as a powerful tool for the understanding of changes in composition.

EXAMPLE. A base iron containing 0.1% phosphorus is to be refined to 0.04% by using a basic oxidizing slag. What slag weight will be required per 100 lb of metal, assuming that the slag-metal equilibrium will result in 6% $(CaO)_4P_2O_5$ in the slag?

If there is no phosphorus in the slag-making constituents to begin with, that is, if we assume that the lime, etc., is free of phosphorus, and since no phosphorus escapes as gas we may write the following mass balance:

P in metal at start = P in metal after treatment with slag

+ P in slag after exposure to metal.

To visualize the problem, it is convenient to begin with an arbitrary weight of metal, say 100 lb. Then

weight of P in 100 lb of ingoing metal = 0.10 lb,

weight of P in 100 lb of outgoing metal = 0.04 lb,

weight of P in slag = 0.06 lb.

Now the phosphorus in the slag is present as $(CaO)_4P_2O_5$, which contains

$$\frac{62}{366} \times 100 = 17\% \text{ P.}$$

TABLE 13–1

	%C	%Mn	%P	%S	%Si
Steel punchings	0.10	0.10	0.03	0.03	0.02
Malleable pig iron	4.0	0.50	0.106	0.07	1.0
Foundry pig iron	4.0	0.50	0.15	0.15	1.5
50% Si ferrosilicon					50.0 balance Fe
80% Mn ferromanganese		80.0			
Analysis:					
Desired	1.5	0.80	0.05	0.05	0.50
			max	max	
Allowed	1.6	0.9	—	—	0.65

Therefore 0.06 lb of phosphorus would be present in 0.354 lb of $(CaO)_4$ P_2O_5 in the slag. Since it is stated in the problem that, under the particular conditions of the experiment, the phosphorus partition results in 6% $(CaO)_4P_2O_5$ in the slag, it follows that the total slag weight will be 0.354/0.06 or 5.9 lb per 100 lb of metal. In other words, a batch process such as an arc furnace would be operating with a slag weight equal to 5.9% of the metal weight.

The familiar charge calculation is another example of a mass balance. In a typical problem concerning an induction-furnace melt, assume that the raw materials listed in Table 13–1 are available to produce an iron of the desired composition. We proceed as follows. We first estimate the melting losses from previous experience, as shown above, and then use these estimates to calculate the allowed analysis. It is then apparent that malleable pig iron, rather than foundry pig iron, should be used as a source of carbon, because of the phosphorus and sulfur limitations.

At this point the inclination would be to set up a group of simultaneous equations to calculate the relative amounts of the malleable pig iron and steel punchings for the charge, but experience shows that it is far easier to use successive estimations. For example, the carbon levels indicate that the majority of the carbon will come from the pig iron and that approximately 40% of the charge must consist of pig iron in order to achieve the desired carbon content of the final metal. Therefore, a first estimate is made to obtain about 1.55%C (or 3.10 lb in a 200-lb charge) from the pig iron, as shown in the calculation below. Letting $x =$ lb pig iron required, we write

$$x \cdot \frac{0.04 \text{ lb carbon}}{1 \text{ lb pig iron}} = 3.1 \text{ lb carbon}, \qquad x = 7.6 \text{ lb pig iron}.$$

TABLE 13–2

SAMPLE CHARGE CALCULATION

Material	lb	C %	C lb	Mn %	Mn lb	P %	P lb	S %	S lb	Si %	Si lb
Malleable pig iron	77.6	4.0	3.10	0.50	0.39	0.06	0.04	0.07	0.06	1.0	0.78
80% Fe Mn	1.62			80	1.30						
50% Fe Si	1.00									50	0.50
Steel punchings	119.78	0.10	0.12	0.10	0.12	0.03	0.04	0.03	0.04	0.02	0.02
First calculation	200.00		3.22		1.81		0.08		0.10		1.30
Allowed composition		1.6	3.2	0.90	1.80	0.05	0.10	0.05	0.10	0.65	1.30
Desired composition		1.5	3.0	0.80	1.60	0.05	0.10	0.05	0.10	0.50	1.00

The Mn, P, S, and Si included in this pig iron are then calculated as shown in the first line of the sample charge calculation presented in Table 13-2. The next step is to calculate the ferroalloys necessary to produce the required Mn and Si contents. It has been estimated that about 60% of the charge will be steel; hence about 0.12 lb of manganese will be obtained from this source, which when added to the quantity of 0.39 lb of manganese from the pig iron leaves a balance of 1.30 lb. The required ferromanganese is calculated next, as shown. The 50% ferrosilicon is similarly computed.

Since the total charge is now 88.22 lb, the balance is made up with steel (119.78). The two-decimal accuracy of the calculation is not significant in the case of steel, but it is important to carry out the weights of ferroalloys to 0.01 lb since 0.02 lb of a 50% alloy is equal to 0.01% element in the final composition.

The elements coming from the steel in the charge are now calculated and the total weights checked with the allowed weights. The weight of carbon is within 0.02 lb, and that of manganese within 0.01 lb of the required amounts, and the silicon weight is correct. The phosphorus is slightly below the maximum and sulfur is at the maximum. The first estimate would be considered satisfactory. Correction for carbon could be accomplished by reducing the quantity of pig iron and replacing it with an equivalent amount of steel. This would have only a second-order effect upon other elements, and no recalculation would be necessary. Manganese could be corrected by increasing ferromanganese and removing the same weight of steel.

In general, the principle followed in these calculations is to compute last the component with lowest alloy content—in our case this is the steel.

PROBLEM 1. Using the same raw materials as listed for the sample charge above, calculate a 400-lb cupola charge, which is to produce metal of the following composition:

C: 3.00, Mn: 0.80, P: 0.20 max, S: 0.13 max, Si: 2.50.

Assume that the effects of cupola melting on composition of the final metal will be as follows:

Mn: − 0.15; P: + 0.01; S: + 0.02 (from coke); Si: − 0.25.

The carbon will increase according to the formula given in Chapter 12. Assume a differential of about $5 per ton in favor of foundry-grade pig.

PROBLEM 2. Consider that the specification now also requires a nickel content of 1% and let us assume that the nickel does not incur any melt-

TABLE 13-3

HEATS OF FORMATION* AT 25°C (77°F) OF CERTAIN COMPOUNDS FROM THEIR ELEMENTS [2]

Compound	Heat of formation,* $-\Delta H$, kcal/mole[†]	Compound	Heat of formation,* $-\Delta H$, kcal/mole[†]
		Oxides	
Al_2O_3	399.0	MnO	92.04
As_2O_3	154.1	MnO_2	124.6
BaO	133.4	Mn_2O_3	232.0
BeO	147.3	Mn_3O_4	331.7
B_2O_3	335.8	MoO_2	130.0
CaO	151.7	MoO_3	176.5
Co (g)	26.416	NiO	58.4
Co_2 (g)	94.052	PbO	52.46
CoO	57.5	P_2O_5	360.0
Cr_2O_3	268.9	SiO_2 (quartz)	206.0
CuO	38.5	SiO_2 (cristobalite)	205.6
Cu_2O	42.5	SnO_2	138.0
FeO	64.6	So_2 (g)	70.96
Fe_2O_3	195.2	So_3 (g)	92.83
Fe_3O_4	266.8	TiO_2	219
H_2O (l)	68.318	V_2O_3	296
H_2O (g)	57.798	WO_2	130.5
MgO	143.8	ZrO_2	258.8
		ZnO	83.4
		Sulfides	
Al_2S_3	121.55	H_2S (g)	4.81
BaS	111.2	MgS	84.39
CaS	113.4	MnS	48.7
CoS	24.18	MoS_2	55.92
CuS	11.6	NiS	20.4
Cu_2S	18.5	S_2 (g)	−31.12
FeS	22.8		
		Miscellaneous	
Al_4C_3	40	NH_3 (g)	11.04
AlN	64		
CaC_2	14.5	SiC	26.7
$Ca(CN)_2$	45.5	TiC	110
$CaCN_2$	83.8	WC	3.9
$CaCO_3$	288.04	Fe_3P	39.1
CaH_2	45.6	$CaO + CO_2 = CaCO_3$	42.31
Ca_3N_2	108.2	$3CaO + Al_2O_3 = 3CaO \cdot Al_2O_3$	2.0
$Ca_3(PO_4)_2$	972.4	$CaO + SiO_2$ (quartz) $= CaSiO_3$	20.7 (800°C)
$CaSi_2$	220	$2CaO + SiO_2$ (quartz) $= Ca_2SiO_4$ (α)	34.27
$CaSO_4$	338.48	$3CaO + SiO_2$ (quartz) $= Ca_3SiO_5$	28.7
CH_4 (g)	17.89	$3CaO + P_2O_5 = Ca_3P_2O_8$	164
C_2H_2 (g)	−54.19	$4CaO + P_2O_5 = Ca_4P_2O_9$	165
CrN	28.26	$3CaO + Al_2O_3 + 2SiO_2 = Ca_3Al_2Si_2O_{10}$	38.7
Cr_4C	16.37	$2FeO + SiO_2 = Fe_2SiO_4$	11.3
Cr_3C_2	21.0	$MnO + SiO_2 = MnSiO_3$	15.0
Fe_3C (α)	−5.8	$2MgO + SiO_2 = Mg_2SiO_4$	33 (1250°C) = 5
Fe_4N	−2.55	$MgO + CO_2 = MgCO_3$	28.27
FeSi	19.2	$3MgO + P_2O_5 = Mg_3P_2O_8$	114.9
Mn_3C	3.6		

* The heat of formation of each element is taken as zero in its normal state at room temperature. These states include alpha iron, rhombic sulfur, red phosphorus, graphitic carbon, gaseous oxygen, nitrogen, and hydrogen. The data apply to compounds in their stable solid state except as indicated.
† To convert to British thermal units per pound mole multiply by 1800.

ing loss. At what cost/ton differential between 3%-nickel steel punchings and plain carbon steel will it become profitable to use the nickel-steel punchings instead of electronickel squares at $0.72/lb?

Energy balances. Mass balances remain constant, and so does the energy content of an isolated system (except in the special case of conversion of mass to energy). It is possible that energy may change in form, as in the conversion from chemical energy to heat energy, but the total energy is unchanged. We now review energy calculations.

We are primarily interested in energy changes, and hence it does not matter if we do not know the total energy of a given substance. If, for example, we produce a reaction between given quantities of coke and oxygen in a bomb calorimeter, we can determine the change in energy between the initial and final states, which, in this case, is the heat liberated:* $\Delta E = q$.

If the reaction were conducted in a cylinder-piston combination doing work w on the surroundings, we would have

$$\Delta E = q - w.$$

If electrical energy resulted as well, another term, w', would be included, and so on. Remember that energy has not been added to the system in any of these cases; the change is a result of the *conversion* of chemical energy to thermal, mechanical, and electrical energy.

In metallurgical reactions we do not usually deal with marked pressure changes. It is simpler to use the concept of change in enthalpy (heat content), ΔH, at constant pressure: $\Delta H = \Delta E + \Delta(pV)$. Note that ΔH is less than ΔE only when work is done by the substance on its surroundings.

Since we cannot measure the absolute value of E or H but must deal with energy changes, the heat of formation of each element in some specified pure state is taken as zero at 25°C (77°F). From the principle of conservation of energy, the heats of formation are algebraically additive. Let us consider the calcining of limestone in the cupola stack, where the reaction is expressed by

$$CaCO_3 = CaO + CO_2.$$

Calorimetric measurements then demonstrate that

$$Ca + \tfrac{1}{2}O_2 = CaO,$$

$\Delta H = -151,700$ cal/mole at 25°C (77°F) and atmospheric pressure
(see Table 13–3).

---

* With respect to the system, the heat q is regarded as positive when heat is absorbed from the surroundings, and as negative when heat is liberated to the surroundings.

TABLE 13–4

SELECTED HEAT-CAPACITY EQUATIONS* [2]

Substance	a	$b \cdot 10^3$	$c \cdot 10^{-5}$	Heat absorbed, cal/mole at indicated temperature °K
Al (s)	4.80	3.22		Fusion 2550 cal at 931.7
Al (l)	7.0			
Al_2O_3 (s)	27.43	3.06	8.47	
CaO (s)	10.00	4.84	1.08	(273 to 1200)
$CaCO_3$ (s)	19.68	11.89	3.08	(273 to 1200)
C (graphite)	2.93	2.0	1.10	
CO (g)	6.60	1.20		
CO_2 (g)	10.50	2.40	2.00	
CH_4 (g)	6.73	10.2	1.12	(273 to 1200)
Cr (s)	5.84	2.36	0.88	
Cr_2O_3 (s)	28.53	2.20	3.74	
Cu (s)	5.44	1.46		Fusion 3110 cal at 1357
Cu (l)	7.50			
H_2 (g)	6.62	0.81		
H_2O (l)	18.03			(273 to 373)
H_2O (g)	7.16	2.58		
H_2S (g)	7.15	3.32		
Fe (average)†	6.12	3.20		Fusion 3670 cal at 1808
Fe (l)	10.5			
FeO (wüstite)	12.08	2.10		Fusion 8600 cal at 1643
Fe_2O_3 (s)	24.72	16.04	4.23	(273 to 1097)
Fe_3O (s)	41.17	18.82	9.80	(273 to 1065)
$FeO \cdot Cr_2O_3$ (s)	38.96	5.34	7.62	
FeS (α)	2.03	39.0		$\alpha \rightarrow \beta$ 1050 cal at 411
FeS (β)	12.05	2.73		Fusion 4670 cal at 1463
MgO (s)	10.86	1.20	2.09	
Mn (average)†	5.12	5.20		
MnO (s)	10.79	2.13	0.56	
Mo (s)	5.69	1.88	0.50	
N_2 (g)	6.50	1.00		
NH_3 (g)	7.12	6.09	0.40	(273 to 1400)
Ni (α)	4.26	6.40		$\alpha \rightarrow \beta$ 92 cal at 626
Ni (β)	6.99	0.91		Fusion 4200 cal at 1725
Ni (l)	8.55			
O_2 (g)	8.27	0.26	1.88	
Si (s)	5.74	0.62	1.01	
SiO_2 (α cristobalite)	3.65	24.0		$\alpha \rightarrow \beta$ 190 cal at 523
SiO_2 (β cristobalite)	17.09	0.45	8.97	
SiC (s)	8.89	2.91	2.84	
S_2 (g)	7.75	0.89		
SO_2 (g)	11.40	1.41	2.04	
TiO_2	17.14	0.98	3.50	
TiC (s)	11.83	0.80	3.58	
V (s)	4.80	2.68	0.60	
V_2O_3 (s)	29.35	4.76	5.42	

* Values of the constants in the equation $C_p = a + bT - cT^{-2}$ for molar heat capacity are expressed in calories per degree at T°K in the range 273 to 1873°K unless otherwise indicated [1].

† Average values for solid from room temperature to melting point, including heats of transformation.

The negative sign shows that heat is evolved when the reaction proceeds from left to right. According to the principle of conservation of energy, the above expression could also be written as

$$CaO = Ca + \tfrac{1}{2}O_2,$$

$$\Delta H = +151{,}700 \text{ cal/mole}.$$

In other words, if a mole of CaO is reduced (as by electrolysis of a fused mass), an energy input equivalent to 151,700 cal is required (at 25°C).

From Table 13–3 the heats of formation of CO_2 and $CaCO_3$ are $-94{,}052$ and $-288{,}040$ cal/mole, respectively, from which we develop the relation for the decomposition of $CaCO_3$:

$$CaCO_3 = CaO + CO_2,$$

$$\Delta H = \Delta H \text{ products } - \Delta H \text{ reactants},$$

$$\Delta H = -151{,}700 - 94{,}052 - (-288{,}040) = 42{,}288 \text{ cal/mole}.$$

In other words, the conversion of $CaCO_3$ to CaO and CO_2 at 25°C requires 42,258 cal/mole.

It is particularly important to note that the above equation does not predict that the reaction *will* take place, but merely describes the energy difference between products and reactants. As will be demonstrated in the next section, it is necessary to heat $CaCO_3$ to above 1623°F (884°C) to cause the reaction to proceed at 1 atm CO_2. As an example, let us calculate ΔH at a fast reaction temperature, 1000°C (1273°K, or 1830°F).

We can develop the proper heat balance by determining the heat required to bring $CaCO_3$ and its products to 1273°K. The difference, ΔC_p, in heat capacities of the products and the reactants, where C_p is the heat capacity per mole, will govern the difference between $\Delta H_{1273°K}$ and $\Delta H_{298°K}$. If we designate by ΔH_1 the heat of reaction at 25°C (77°F) and by ΔH_2 that at 1000°C, then

$$\Delta H_2 - \Delta H_1 = \int_{T_1}^{T_2} \Delta C_p \, dT.$$

If the heat capacity for each substance is expressed by an empirical equation of the form

$$C_p = a + bT - cT^{-2},$$

where a, b, and c are constants for a particular substance, the algebraic sum of the heat capacities of the products minus that of the reactants will be

$$\Delta C_p = \Delta a + \Delta b T - \Delta c T^{-2}$$

and

$$\Delta H_2 - \Delta H_1 = \int_{T_1}^{T_2} \Delta C_p \, dT,$$

or

$$\Delta H = \Delta H_0 + \Delta a T + \tfrac{1}{2} \Delta b T^2 + \Delta c T^{-1}. \tag{1}$$

Although ΔH_0 is used here merely as a constant of integration, it could be considered to be ΔH at absolute zero if the heat capacity values could be extrapolated to this point. From Table 13–4 (p. 294) we have the following heat capacity equations:

$$C_p(CaO) = 10.00 + 4.84 \times 10^{-3} T - 1.08 \times 10^{+5} T^{-2}$$

$$+ C_p(CO_2) = 10.50 + 2.4 \times 10^{-3} T - 2.00 \times 10^{+5} T^{-2}$$

$$- C_p(CaCO_3) = -19.68 - 11.89 \times 10^{-3} T + 3.08 \times 10^{+5} T^{-2}$$

$$\overline{\Delta C_p = +0.82 - 4.65 \times 10^{-3} T - 0}$$

From the above, we obtain:

$$\Delta a = 0.82,$$

$$\Delta b = -4.65 \times 10^{-3},$$

$$\Delta c = 0,$$

and substituting these values in the ΔH-equation yields

$$\Delta H = \Delta H_0 + 0.82 T - 2.33 T^2 \times 10^{-3}.$$

Since we know $\Delta H = 42{,}288$ cal at $25°C$ ($298°K$), ΔH_0 is found to be $42{,}258$ cal/mole, and therefore $\Delta H_{1273°K} = 39{,}498$ cal/mole.

The same answer may be obtained by using the ΔH-equations provided by Kelley [1]. As an example, the enthalpy of CO_2 is expressed by the following equation:

$$H_T - H_{298.16} = 10.55\,T + 1.08 \times 10^{-3} T^2 + 2.04 \times 10^5 T^{-1} - 3926.$$

13–5 Combined application of the first and second laws of thermodynamics. While ΔH-balances are valuable, they do not predict whether a reaction will occur, how temperature will affect the reaction, and what the balance of products and reactants will be at equilibrium. We know qualitatively that the addition of manganese to a bath of oxidized steel will cause the reaction $[Mn] + [O] = (MnO)_{slag}$, but it is important to know also the amount of $[O]$ remaining as well as the effect of the temperature upon the reaction. For this calculation we need to define a new function, the free energy F, which is a measure of the driving force of a reaction.

The driving force of a reaction is dictated by the amount of available energy. We know, for example, that heat cannot be completely converted to work, even in an ideal frictionless engine. We must, therefore, deduct from ΔH (the heat of reaction) a quantity $T\Delta S$ which is a measure of the unavailable energy. Then ΔF is the maximum energy available for electrical, mechanical, or other work:

$$\Delta F = \Delta H - T\Delta S,$$

where ΔS is expressed in cal/mole°K and is called the entropy change. The quantity $T\Delta S$ has the same units as ΔH (cal/mole). If the free energy of a given reaction decreases (ΔF is negative), the reaction tends to occur. If ΔF is positive, the reaction does not occur; in fact, it tends to go in the *opposite* direction.

We have already discussed the calculation of ΔH at any temperature; to complete the free-energy calculation we need only the value for ΔS (Table 13–5) and its variation with temperature [1]. As an example, let us assume that we wish to explore the temperature range in which limestone calcines to CaO and CO_2 in the cupola stack. For the reaction

$$CaCO_3(s) = CaO(S) + CO_2(g),$$

we have already determined that

$$\Delta H = 42{,}258 + 0.82T - 2.33 \times 10^{-3}T^2.$$

As an exploratory calculation, let us consider that at a temperature of $1000°C = 1832°F = 1273°K$,

$$\Delta H = 39{,}498 \text{ cal/mole.}$$

To evaluate $T\Delta S$, we find, from Table 13–5:

	$CaCO_3(s)$	$= CaO(s)$	$+ CO_2(g)$
$S_{25°C}$	22.2	9.5	51.1 cal/deg-mole
$S_{1000°C} - S_{25°C}$*	40.3	19.4	17.6
$S_{1000°C}$	62.5	$=$ 28.9	68.7
ΔS		$=$ 35.1 cal/deg-mole	
$T\Delta S$		$= (1273)(35.1) = 44{,}700$ cal/mole	

*These values are from Kelley [1]. For other calculations, the required entropies are already included in the important equations of Tables 13–6, 13–7, and 13–8. They also can be calculated directly from the expression

$$S_T = \int_0^T (C_p/T)dT$$

in an analogous way to ΔH.

TABLE 13–5

ENTROPIES OF SUBSTANCES AT 25°C (77°F) [2]

Substance	Entropy, cal/deg-mole	Substance	Entropy, cal/deg-mole
Al	6.77 ± 0.05	Cu	7.97 ± 0.05
Al$_2$O$_3$	12.5 ± 0.15	Cu$_2$O	24.1 ± 1.5
As	8.4 ± 0.2	CuO	10.4 ± 0.2
As$_2$O	25.6 ± 0.5	Cu$_2$S	28.9 ± 0.8
Be	2.28	CuS	15.9 ± 0.4
BeO	3.37	H$_2$	31.23 ± 0.01
B	1.7 ± 0.2	H	27.40 ± 0.01
B$_2$O$_3$	13.0 ± 0.1	H$_2$O (l)	16.57 ± 0.03
Ca	9.95 ± 0.1	H$_2$O (g)	45.13 ± 0.01
Ca (g)	37.00 ± 0.01	H$_2$S	49.15 ± 0.1
CaCO$_3$ (calcite)	22.2 ± 0.2	Fe	6.49 ± 0.03
CaC$_2$	16.8 ± 0.5	FeO	14.2 ± 2.0
CaF$_2$	16.4 ± 0.4	Fe$_2$O$_3$	21.5 ± 0.5
CaH$_2$	9.9 ± 1.0	Fe$_3$O$_4$	35.0 ± 0.7
CaO	9.5 ± 0.2	FeS	16.1 ± 0.3
CaS	13.5 ± 0.3	FeS$_2$	12.7 ± 0.2
CaSO$_4$	25.6 ± 0.3	Pb	15.51 ± 0.05
C (diamond)	0.6 ± nil	PbO	16.6 ± 0.4
C (graphite)	1.36 ± 0.03	PbS	21.8 ± 0.6
CH$_4$	44.46 ± 0.1	Mg	7.77 ± 0.1
CO	47.32 ± 0.01	Mg (g)	35.51 ± 0.01
CO$_2$	51.08 ± 0.1	MgO	6.55 ± 0.1
CS$_2$ (g)	56.84 ± 0.3	Mg$_2$SiO$_4$	22.7 ± 0.2
Cr	5.68 ± 0.07	Mn	7.59 ± 0.04
Cr$_2$O$_2$	19.4 ± 0.3	MnO	14.4 ± 0.6
MnS	18.7 ± 0.3	SO$_2$ (g)	59.40 ± 0.2
N$_2$	45.79 ± 0.01	Sn	12.3 ± 0.1
NH$_2$	46.03 ± 0.1	SnO	13.5 ± 0.3
O$_2$	49.03 ± 0.01	SnO$_2$	12.5 ± 0.3
Si	4.50 ± 0.05	Ti	7.24 ± 0.07
SiO	3.95 ± 0.04	TiC	5.8 ± 0.1
SiO$_2$ (quartz)	10.1 ± 0.1	TiO$_2$	12.01 ± 0.05
SiO$_2$ (cristobalite)	10.35 ± 0.1	V	7.0 ± 0.1
S (rhombic)	7.62 ± 0.1	V$_2$O$_3$	23.5 ± 0.3
S$_2$ (g)	54.41 ± 0.1	ZrO$_2$	12.03 ± 0.08

Therefore

$$\Delta F = 39,498 - 44,700,$$

$$\Delta F = -5,200 \text{ cal/mole.}$$

Since ΔF is negative, the reaction can occur.

If we continue this calculation to lower temperatures, we find that $\Delta F = 0$ at 884°C (1623°F) and that below this temperature ΔF is positive; hence the reaction will not occur spontaneously under the conditions at hand, i.e. at 1 atm pressure of CO_2. This leads us to infer that ΔF is not only a quantitative measure of whether a reaction can or cannot occur, but it also governs the ratio of reactants and products.

Let us now see how this relationship applies in the deoxidation of steel with silicon. We have the reaction

$$(SiO_2)_{slag} = [Si]_{metal} + 2[O]_{metal}.$$

At equilibrium, the silicon and oxygen dissolved in liquid steel are in equilibrium with SiO_2 slag. Then

$$K_1 = \frac{a_{Si} \times a_O^2}{a_{SiO_2}},$$

and since in heterogeneous equilibria the activity of a pure phase is taken to be unity, we have

$$K_2 = a_{Si} \times a_{[O]}^2.$$

At low concentrations, a_{Si} and a_O are proportional to the weight percentages and we may therefore write

$$K = \%[Si] \times \%[O^2].$$

If we can now relate K to free energy changes, we will be able to *calculate* the %[O] at any %[Si] in deoxidation and compare the effectiveness of Si with Mn, Al, and other deoxidants.

From physical chemistry we have the relation

$$\Delta F° = -RT \ln K,$$

where $\Delta F°$ represents the free energy of the products of a reaction in their *standard states* minus the free energy of the reactants *in their standard states*.* Values of $\Delta F°$ for important steelmaking reactions are given in Tables 13–6, 13–7, and 13–8. Therefore, to find K for the re-

---

* In these tables the standard state is defined so that the activity of the element in solution is equal to its percentage by weight.

TABLE 13-6

ENTHALPY AND ENTROPY CHANGES IN GAS REACTIONS FOR COM-
PUTATION OF STANDARD FREE-ENERGY CHANGE AT 1600°C (1873°K) [2]

$$\Delta F^\circ = \Delta H - T\Delta S$$

No.	Reaction	ΔH, cal/mole	ΔS, cal/mole/°K
1	H_2 (g) $+$ 1/2 O_2 (g) $=$ H_2O (g)	$-$ 60,180	-13.93
2	CO (g) $+$ 1/2 O_2 (g) $=$ CO_2 (g)	$-$ 66,560	-20.15
3	C (gr) $+$ 1/2 O_2 (g) $=$ CO (g)	$-$ 28,100	$+20.20$
4	C (gr) $+$ C_2 (g) $=$ CO_2 (g)	$-$ 94,640	$+$ 0.05
5	H_2 (g) $+$ CO_2 (g) $=$ H_2O (g) $+$ CO (g)	$+$ 6,380	$+$ 6.22
6	C (gr) $+$ CO_2 (g) $=$ 2CO (g)	$+$ 38,460	$+40.35$
7	H_2 (g) $+$ 1/2 S_2 (g) $=$ H_2S (g)	$-$ 21,680	-11.81
8	1/2 S_2 (g) $+$ O_2 (g) $=$ SO_2 (g)	$-$ 86,380	-17.30
9	1/2 S_2 (g) $+$ 1/2 O_2 (g) $=$ SO (g)	$-$ 6,720	$+$ 1.25
10	C (gr) $+$ S_2 (g) $=$ CS_2 (g)	$-$ 3,600	$+$ 1.44
11	C (gr) $+$ 2H_2 (g) $=$ CH_4 (g)	$-$ 21,960	-26.61
12	2C (gr) $+$ H_2 (g) $=$ C_2H_2 (g)	$+$ 53,200	$+12.66$
13	H_2 (g) $=$ 2H (g)	$+108,300$	$+28.80$

action at 1600°C (2912°F), we calculate

$$(SiO_2)_{slag} = [Si]_{metal} + 2[O]_{metal}$$

$$\Delta F^\circ = +38,840 \text{ cal/mole},$$

$$\log_{10} K = - \frac{\Delta F^\circ}{4.575T},^*$$

$$\log K_{1600°C} = - \frac{38,840}{8590} = -4.53 = 5.47 - 10,$$

$$K_{1600°C} = 3.0 \times 10^{-5} = \% \text{ [Si]} \times \% \text{ [O]}^2.$$

The variation of K with temperature is of considerable interest, since
it enables us to select the optimum temperature for a given reaction. In
the case of the silicon deoxidation just discussed, we have from Table
13-8:

$$\log K = - \frac{\Delta F^\circ}{4.575T}$$

$$= - \frac{129,440 - 48.44T}{4.575T}$$

$$= - \frac{28200}{T} - 10.55.$$

---

* The value 4.575 arises from insertion of the value of 1.986 cal/deg for K and
also changing to $\log_{10}$ from ln base e (1.986 $\times$ 2.303).

TABLE 13–7

ENTHALPY AND ENTROPY CHANGES IN FORMATION OF OXIDES, SULIFIDES, NITRIDES, AND CARBIDES FOR COMPUTATION OF STANDARD FREE-ENERGY CHANGE AT STEELMAKING TEMPERATURES [2]

No.	Reaction	ΔH, cal/mole	ΔS, cal/mole/°K
	Oxides		
1	Fe (l) $+$ 1/2 O_2 $=$ FeO (l)	$-$ 56,830	$-$11.94
2	Mn (l) $+$ 1/2 O_2 $=$ MnO (s)	$-$ 97,000	$-$21.4
3	MnO (s) $=$ MnO (l)	$+$ 10,700	$+$ 5.2
4	Ni (l) $+$ 1/2 O_2 $=$ NiO (s)	$-$ 60,750	$-$25.1
5	Co (l) $+$ 1/2 O_2 $+$ CoO (s)	$-$ 60,530	$-$19.6
6	Be (l) $+$ 1/2 O_2 $=$ BeO (s)	$-$146,730	$-$22.1
7	Mg (g) $+$ 1/2 O_2 $=$ MgO (s)	$-$176,500	$-$47.5
8	Ca (g) $+$ 1/2 O_2 $=$ CaO (s)	$-$192,000	$-$49.1
9	2Al (l) $+$ 3/2 O_2 $=$ Al_2O_3 (s)	$-$400,000	$-$76.6
10	2Cr (s) $+$ 3/2 O_2 $=$ Cr_2O_3 (s)	$-$265,050	$-$60.4
11	2V (s) $+$ 3/2 O_2 $=$ V_2O_3 (s)	$-$287,300	$-$54.3
12	Si (l) $+$ O_2 $=$ SiO_2 (s)	$-$214,300	$-$47.0
13	Ti (s) $+$ O_2 $=$ TiO_2 (s)	$-$217,600	$-$41.9
14	Zr (s) $+$ O_2 $=$ ZrO_2 (s)	$-$256,000	$-$44.0
15	Mo (s) $+$ O_2 $=$ MoO_2 (s)	$-$137,000	$-$39.4
16	W (s) $+$ O_2 $+$ WO_2 (s)	$-$139,150	$-$41.7
	Sulfides		
17	Fe (l) $+$ 1/2 S_2 (g) $=$ FeS (l)	$-$ 34,000	$-$10.4
18	Mn (l) $+$ 1/2 S_2 (g) $=$ MnS (s)	$-$ 68,700	$-$19.1
19	Mg (g) $+$ 1/2 S_2 (g) $=$ MgS (s)	$-$132,300	$-$45.7
20	Ca (g) $+$ 1/2 S_2 (g) $=$ CaS (s)·	$-$169,600	$-$47.4
21	2Cu (l) $+$ 1/2 S_2 (g) $=$ Cu_2S (l)	$-$ 29,300	$-$ 6.2
22	Mo (s) $+$ S_2 (g) $=$ MoS_2 (s)	$-$ 76,300	$-$33.3
23	W (s) $+$ S_2 (g) $=$ WS_2 (s)	$-$ 69,800	$-$32.9
24	Si (l) $+$ S_2 (g) $=$ SiS_2 (s)	$-$ 73,200	$-$44.0
25	2Al (l) $+$ 3/2 S_2 (g) $=$ Al_2S_3 (s)	$-$164,400	$-$69.0
	Nitrides		
26	3Be $+$ N_2 $=$ Be_3N_2 (s)	$-$133,500	$-$40.6
27	B (s) $+$ 1/2 N_2 $=$ BN (s)	$-$ 27,700	$-$10.4
28	Al (l) $+$ 1/2 N_2 $=$ AlN (s)	$-$ 62,300	$-$30.1
29	Ti (s) $+$ 1/2 N_2 $=$ TiN (s)	$-$ 80,300	$-$21.0
30	V (s) $+$ 1/2 N_2 $=$ VN (s)	$-$ 43,000	$-$21.4
31	Zr (s) $+$ 1/2 N_2 $=$ ZrN (s)	$-$ 82,200	$-$22.0
32	3Si (l) $+$ 2N_2 $=$ Si_3N_4 (s)	$-$208,600	$-$97.2
	Carbides		
33	Ca (g) $+$ 2C (gr) $=$ CaC_2 (s)	$-$ 59,800	$-$21.6
34	4Al (l) $+$ 3C (gr) $=$ Al_4C_3 (s)	$-$ 35,700	$-$28.1
35	3Cr (s) $+$ 2C (gr) $=$ Cr_3C_2 (s)	$-$ 8,550	$+$ 5.0
36	Si (l) $+$ C (gr) $=$ SiC (s)	$-$ 38,400	$-$ 8.5
37	Ti (s) $+$ C (gr) $=$ TiC (s)	$-$ 57,300	$-$ 2.5
38	2Mo (s) $+$ C (gr) $=$ Mo_2C (s)	$+$ 4,200	$+$ 4.8

TABLE 13–8

STANDARD FREE ENERGY OF SOLUTION OF VARIOUS
ELEMENTS IN LIQUID IRON* [2]

Element, state	$\gamma°_{1873}$	$\Delta F°$
Nickel (l)	1	$-\ 9.21T$
Manganese (l)	1	$-\ 9.11T$
Cobalt (l)	1	$-\ 9.26T$
Chromium (s)	1	$4{,}350 - 11.11T$
Molybdenum (s)	1	$6{,}280 - 12.32T$
Tungsten (s)	1	$7{,}640 - 13.62T$
Silicon (l)	0.0178	$-\ 29{,}000 -\ 0.30T$
Copper (l)	12	$+\ 9{,}300 -\ 9.4T$
Vanadium (s)	0.1	$-\ 3{,}900 - 11.07T$
Aluminum (l)	0.043	$-\ 11{,}700 -\ 7.7T$
Titanium (s)	(0.05)	$-\ 7{,}000 - 11.0T$
Zirconium (s)	(0.05)	$-\ 7{,}000 - 12.2T$
Carbon (graphite)		$+\ 8{,}900 - 12.10T$
Nitrogen $\frac{1}{2}N_2$ (g) $=$ [N]		$+\ 2{,}580 +\ 5.02T$
Hydrogen $\frac{1}{2}H_2$ (g) $=$ [H]		$+\ 7{,}640 +\ 7.68T$
Oxygen $\frac{1}{2}O_2$ (g) $=$ [O]		$-\ 27{,}930 -\ 0.57T$
Oxygen FeO (l) $=$ [O] $+$ Fe		$+\ 28{,}900 - 12.51T$
Sulfur $\frac{1}{2}S_2$ (g) $=$ [S]		$-\ 31{,}520 +\ 5.27T$
CO_2 (g) $+$ C (gr) $=$ 2CO (g)		$+\ 38{,}460 - 40.35T$
CO_2 (g) $+$ [C] $=$ 2CO (g)		$+\ 29{,}550 - 28.24T$
H_2 (g) $+$ [O] $=$ H_2O (g)		$-\ 32{,}250 + 14.50T$
H_2O (g) $=$ 2[H] $+$ [O]		$+\ 47{,}530 +\ 0.86T$
CO (g) $+$ [O] $=$ CO_2 (g)		$-\ 38{,}050 + 20.72T$
Fe (l) $+$ 1/2 O_2 (g) $=$ FeO (l)		$-\ 56{,}830 + 11.94T$
[C] $+$ [O] $=$ CO (g)		$-\ 8{,}510 -\ 7.52T$
MnO (l in FeO) $=$ [Mn] $+$ [O]		$+\ 58{,}400 - 25.98T$
SiO_2 (s) $=$ [Si] $+$ 2[O]		$+129{,}440 - 48.44T$
Cr_2O_3 (s) $=$ 2[Cr] $+$ 3[O]		$+189{,}960 - 84.33T$
V_2O_3 (s) $=$ 2[V] $+$ 3[O]		$+196{,}600 - 78.1T$
H_2 (g) $+$ [S] $=$ H_2S (g)		$+\ 9{,}840 +\ 6.54T$
[S] $+$ 2[O] $=$ SO_2 (g)		$+\ 1{,}340 + 12.81T$

* Reference state is the infinitely dilute solution in pure liquid iron,
in which the activity of the added element is equal to its percentage
by weight.

The plot of $\log K$ versus $1/T$ is therefore a straight line with negative slope, that is, $\log K$ decreases as $1/T$ increases or as T decreases. Hence the product $[Si] \times [O]^2$ decreases with decreasing temperature, which shows that deoxidation with a given amount of $[Si]$ improves as the temperature is lowered. Conversely, the amount of $[Si]$ that can be reduced from SiO_2 (as in acid-electric practice) increases with temperature if $[O]$ is kept constant.

13–6 Gas-liquid equilibrium. A good deal of the experimental work dealing with equilibrium is concerned with the determination of activity coefficients since, although the chemical analysis of a metal or slag solution gives us each concentration, it is the active concentration (the "activity") which governs the equilibrium constant. The activity coefficient is a correction factor which is applied to the concentration to give the activity. It is expressed in either of two ways: by Raoult's law or by Henry's law.

Raoult's law deals with ideal solutions. For example, a mixture of perfect gases is an ideal gaseous solution, in which the activity of each component is equal to its mol fraction. The partial pressure of each gas is often used as a measure of activity. In an ideal solution of two metals, then, each exerts a partial vapor pressure proportional to its concentration at all concentrations, provided also that the vapor behaves like an ideal gas. However, this is rarely the case in solution. The metal atoms either attract one another, thus lowering the expected pressure, or repel one another, thus raising the pressure. To develop true equilibrium constants, it is necessary to apply activity coefficients which are determined experimentally. This is done by means of Raoult's or Henry's laws. Raoult's law states that in an ideal solution the activity is *equal* to the mol fraction, while Henry's law for dilute solutions merely states that the activity is *proportional* to the mol fraction or, approximately, to the weight percentage.

Raoult's law. The activity $a_i = \gamma_i N_i$, where γ_i is the activity coefficient and N_i is the mol fraction.

Henry's law. At low concentrations (up to 5%) Henry's law is used. Since the weight percent is approximately proportional to the mol fraction of a given metal at low percentages, the law can, for convenience, be stated as follows:

$$a_i = (f_i)(\%i),$$

where f_i is the coefficient so chosen that, usually, $a_i = f_i$ in a 1%-solution.

PROBLEM. The vapor pressure of pure zinc at 1949°F (1065°C) is 4 atm, while that of copper is 10^{-4}. A 10% (atomic percent) solution of zinc in copper exhibits a vapor pressure of 0.15 atm at 1949°F. Calculate

γ_i at this percentage. Are the Zn-Cu forces attractive or repulsive? Would you expect a positive or negative heat of solution of Zn in Cu?

From the following data, calculate f_i and a_i at 1.15 and 2.5% Zn by weight:

Zn, w/o (balance Cu)	p, atm
0.9	0.09
1.1	0.11
1.2	0.12
2.0	0.25
3.0	0.37

Gas-metal equilibria and partial pressures. The relations just discussed lead naturally to considerations of equilibria. It follows that

$$p_i = p_i^0 \cdot a_i,$$

where p_i and p_i^0 are vapor pressures in unknown and in standard states, respectively. If a melt is under atmospheric pressure, gas can only be evolved if the sum of the partial pressures (p_i, p_j, p_k, etc.) is greater than one atmosphere.

Clapeyron equation. The vapor pressure of a pure metal may be calculated from the modified Clapeyron equation:

$$\log p = \frac{-\Delta H}{4.575T} + B,$$

where ΔH is the heat of vaporization and B is a constant for the particular substance. The results of this calculation for the common metals are given in Fig. 13–1.

Nucleation of bubbles. It is often taken for granted that when the sum of the pressures of the constituents in a melt exceeds one atmosphere, a boil will be encountered. However, *nucleation* is necessary. In a spherical gas bubble, the relation between the extra pressure in the bubble which balances the surface tension of the liquid is given by the formula:

$$\Delta p = \frac{2S}{r},$$

where S is the surface tension of the liquid and r is the radius of the bubble. If the surface tension of steel is taken as 1500 dynes/cm, a bubble $\frac{1}{16}$ mm in diameter is subject to an extra pressure of 1 atm from this source. This effect of surface tension explains why a melt may boil upon the insertion of a dry sampling spoon or upon pouring into a dry ladle. The irregular solid surfaces or vortices in the liquid result in lowered surface tension and can nucleate bubbles.

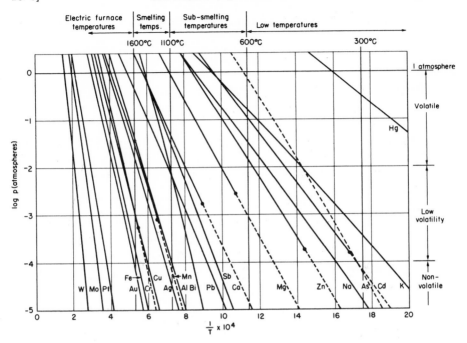

FIG. 13–1. Vapor pressures of common metals. (Solid lines represent data for liquid metals; dashed lines for solid metals.) [3]

Seivert's law. One of the most frequently required calculations involves the question, "If the atmosphere over a melt contains a partial pressure of x atmospheres of gas G, what percentage is dissolved in the melt?" This is really a special case of the law of mass action. Most gases dissolve atomically, but are present as diatomic molecules above the melt. Therefore, we have the reaction

$$G_2 = 2[G],$$

where $[G]$ is the dissolved gas and G_2 is the molecular gas contained in the atmosphere or in the bubbles passing through the bath. If we assume that the activity in the liquid follows Henry's law and is proportional to the weight percentage, and that the gas behaves ideally in the atmosphere and the activity is proportional to its partial pressure, then

$$K = \frac{a_{[G]}^2}{a_{G_2}} = \frac{\%_{[G]}^2}{p_{G_2}}.$$

Now K may be evaluated for a given temperature by substituting a known set of values. For example, the solubility of H_2 in iron at one atmosphere pressure of H_2 over the melt at 2820°F (1549°C) is 0.0025%, i.e.,

$$K = \frac{0.0025^2}{1}.$$

Then with a partial pressure of 0.01 atm H_2 over the melt, we have

$$[H] = \sqrt{(0.0025)^2(0.01)} = 0.00025\%.$$

PROBLEM. Calculate the equilibrium %[H] for an aluminum melt purged with nitrogen that is contaminated with 1% H_2 by volume. Select the optimum practical temperature for this reaction, using the data of Chapter 10.

13–7 Slag-metal equilibria. The concepts of slag-metal equilibria and activity relations have greatly simplified metallurgical calculations. Let us take the example of a cupola which is to be operated under desulfurizing conditions for the production of spherulitic irons. A maximum of 0.02% S is desired. The charge consists of steel, return scrap, coke, and limestone, all of which may contain sulfur. If we try to write a careful description of the process, the details become needlessly involved. The only important results are the ratio $[S]_{slag}/[S]_{metal}$ for these conditions and the weights of slag and metal. If, for example, the ingoing total sulfur amounts to 0.10% and a final $[S]_{metal}$ of 0.02% is desired, we see that a good $[S]_{slag}/[S]_{metal}$ ratio is 80, with a basic reducing slag (Fig. 12–8). The slag-metal weight ratio needed can then be calculated by the method of simple proportions, as in the following problem.

PROBLEM. The ratio $\%[S]_{slag}/\%[S]_{metal}$ between a given slag and a cast iron containing 2.5% Si, 3.8% C is 80.

The charge is composed of 60% steel (S = 0.06%) and 40% scrap (S = 0.02%). A 5 : 1 metal-to-coke ratio is used. The sulfur of the coke is 0.50%, and we assume that the limestone and fluorspar are free of sulfur. What slag/metal ratio by weight should be used?

13–8 Combined phase-diagram and equilibrium-constant calculations. Cu-O-S-SO_2. Although it is important to know the equilibria between gas and metal solutions at refining temperatures, the equilibria during solidification are even more important for an understanding of gas defects. Not only does the equilibrium constant change from what it was at refining temperatures, but the concentration of impurities taking place in the liquid as the relatively pure solid precipitates is a major problem.

The copper-oxygen-sulfur system is an excellent example. The evolution of SO_2 was discussed in Chapter 10, but we review the important calculations here.

The equilibrium constant for the equation $SO_2 = 2[O] + [S]$ is

$$K = \frac{[S] \times [O^2]}{p_{SO_2}}$$

and decreases from 3.3×10^{-5} at $1250°C$ ($2282°F$) to 0.98×10^{-5} at $1083°C$ ($1981°F$), the freezing point of copper, which indicates a reduced tolerance for either [S] or [O]. Even with the smaller equilibrium constant, it would appear that relatively high values of [S] and [O] are called for to produce p_{SO_2} in excess of 160 mm and cause gas bubbles: 0.04% [S] at 0.38% [O], for example. Calculation of various [S] and [O] concentrations yields the curve of Fig. 10–22.

Now let us determine the variation in S and O in the liquid copper portion of the melt under equilibrium conditions as solidification proceeds. Of course, as soon as the product $[S] \times [O^2]$ reaches the equilibrium constant, SO_2 gas will be formed if there are no nucleation difficulties, and no further buildup of S and O will take place. Assuming a starting concentration of 0.02% [S] and 0.040% [O], we obtain the results listed in Table 13–9.

TABLE 13–9

EQUILIBRIA DURING SOLIDIFICATION

(Assuming no solid solubility of sulfur or oxygen)

% Solid	0	50	70	80
% Liquid	100	50	30	20
[S] (l)	0.02	0.04	0.13	0.20
[O] (l)	0.040	0.08	0.27	0.40
$[O^2]$	1.6×10^{-3}	6.4×10^{-3}	7.3×10^{-2}	1.6×10^{-1}
$[S] \times [O^2]$	3.2×10^{-5}	2.6×10^{-4}	9.5×10^{-3}	8.0×10^{-2}
p_{SO_2}, mm	3.27	26.6	950	8190

The values for p_{SO_2} are obtained, using $K = 0.98 \times 10^{-5}$, as follows:

$$p_{SO_2}, \text{mm} = \frac{[S] \times [O^2]}{0.98 \times 10^{-5}}$$

The above calculations demonstrate how impurities may concentrate in the liquid, and give rise to gas evolution.

The role of a residual deoxidant in this calculation is interesting. Phosphorus will maintain O at a low value. The presence of 0.02% P will buffer the oxygen at 0.002% so that the [O]-concentration will not increase from 0.04 to 0.4%, but will remain low; hence the product $[S] \times [O]^2$ will be greatly reduced. A similar effect is obtained with zinc, i.e., 2.5% Zn added to a 10% tin bronze lowers the oxygen to 0.002%.

In steel metallurgy the same effect prevents the evolution of CO if the [Al] content is high enough. Oxygen, which is practically insoluble in solid iron, tends to concentrate in the melt but buffering by Al results in inert alumina inclusions.

REFERENCES

1. K. K. Kelly, "Contributions to the Data on Theoretical Metallurgy: X. High Temperature Heat Content, Heat Capacity and Entropy Data for Inorganic Compounds," *Bulletin,* Bureau of Mines, U. S. Department of Interior, 1949.

2. *Basic Open Hearth Steelmaking,* A.I.M.E., 1952.

3. R. Schumann, *Metallurgical Engineering,* Vol. I. Reading, Massachusetts: Addison-Wesley Publishing Co., 1952.

GENERAL REFERENCES

1. O. Kubachewski and E. L. Evans, *Metallurgical Thermochemistry.* New York: Wiley, 1956.

2. L. S. Darker and R. W. Gurry, *Physical Chemistry of Metals.* New York: McGraw-Hill Book Co., 1953.

APPENDIX I

GLOSSARY OF COMMON FOUNDRY TERMS

Anchor. Appliance used to hold cores in molds.

Arbor. A device to reinforce or lift a mass of sand.

Bedding in. Sinking a pattern into the sand by excavating a "bed" in which the pattern is placed for ramming up.

Binders. Materials used to hold molding sand together.

Blow hole. Hole in the casting caused by trapped air or gases.

Bottom board. The board that the mold rests upon.

Casting. The metal shape, exclusive of gates and risers, that is obtained as a result of pouring metal into a mold.

Chaplet. A metal support used to hold a core in place in a mold. Not used when a core print will serve.

Cheek. The portion of a flask placed between the cope and drag when a mold has more than two sections.

Chill. A metal object placed in the wall of a mold, causing the metal to solidify more rapidly at such a point.

Close over. The operation of lowering a part of the mold over some projecting portion such as a core.

Cold shut. The imperfect junction where two streams of metal meet but do not fuse.

Contraction. Decrease in size due to cooling of the metal after it is poured. Shrinkage is the term applied to the decrease in volume of metal from liquid to solid stage. Contraction immediately follows it.

Cope. The upper or topmost section of a flask.

Core. A separate part of the mold, made of sand and generally baked, which is used to create openings and various shaped cavities in the casting.

Core box. A mold in which a core is formed.

Core dryer. A metal form in which the core is baked.

Core rod. Irons or bars imbedded in a core to strengthen it.

Crushing. The pushing out of shape of core or mold when two parts of the mold do not fit properly.

Dowel. A pin used between the sections of parted patterns or core boxes to locate and hold them in position.

Draft. Slight taper given to pattern to allow drawing from the sand.

Drag. The bottom part of a flask or mold.

Drawback. A part of the mold, made of green sand, which may be drawn back to clear overhanging portions of the pattern. It is rammed up on a plate or arbor so that it can be lifted away.

Drawing. Removing the pattern from the sand.

Drop or drop out. The falling away of a body of sand when the mold is jarred or lifted.

Feeding. Supplying additional molten metal to a casting to make up for volume shrinkage during solidification.

Flask. A metal or wood frame, without fixed top or bottom, in which the mold is formed.

Flask pins. Pins to fit corresponding sockets on the joint of a flask to permit separation.

Feed head. A reservoir of molten metal from which the casting feeds as it solidifies. Also called a riser.

Fin. A thin projection on a casting due to an imperfect joint in the mold.

Fillet. A concave corner piece used at the intersection of two surfaces to round out a sharp corner.

Follow board. A board shaped to the parting line of the mold.

Gate. A channel through which the molten metal enters the casting cavity.

Gaggers. Metal supports shaped like the letter "L" that are used to reinforce the sand in the mold.

Green sand. Molding sand tempered with water to proper consistency for foundry use.

Green-sand core. A core that is made of molding sand but not baked.

Head. The pressure exerted by a column of fluid, such as molten metal, water, etc.

Hot spots. Areas of extra mass usually found at the junction of sections.

Hot tears. Cracks in metal castings formed at elevated temperatures by contraction stresses.

Jarring machine. A molding machine that packs the sand by jarring.

Jig. A device arranged to expedite a hand or machine operation.

Loam mold. A mold built up of brick, covered with a loam mud, and then baked before being poured.

Loose piece. Part of a pattern that remains in the mold and is taken out after the body of the pattern is removed.

Machine finish. Allowance of stock on the surface of the pattern to permit machining of the casting to the required dimensions.

Master pattern. An original pattern made to produce castings which are then used as metal patterns.

√ *Match plate.* A plate to which the pattern is attached at parting line.

Mold. A body of molding sand or other heat-resisting material containing a cavity which forms a casting when filled with molten metal.

Molding sand. A sand which binds strongly without losing its permeability to air or gases.

Nowel. The lower section of the flask; commonly called the drag.

Overhang. The extension on the vertical surface of a core print, providing clearance for closing the mold over the core, also known as "shingle."

Parting. Joint where mold separates to permit removal of pattern.

Parting sand. A bondless sand dusted on the parting to prevent the parts of the mold from adhering to each other.

Pouring. Filling the mold with molten metal.

Ramming up. The process of packing the sand in the mold or core box with a rod or rammer.

Rapping. Loosening the pattern from the mold by jarring or knocking.

Rechucking. Reversing a pattern upon a face plate to permit turning the opposite face to the required shape.

Riser. A column of metal placed in the mold to feed the casting as it shrinks and solidifies. Also known as a "feed head."

Rolling over. Operation of turning flask over to reverse its position.

Runner. The channel through which the molten metal is carried from the sprue to the gate.

Shrinkage. The decrease in volume when molten metal solidifies.

Shrink hole. A cavity in a casting due to insufficient feed metal.

Sizing. A primary coating of glue applied to the end grain of wood to seal the pores.

Skeleton pattern. A framework representing both the exterior and interior shape of the casting.

Slab core. A plain flat core.

Snap flask. A flask that has hinges and latches so that it may be removed from the mold prior to the pouring.

Soldiers. Wooden pegs used to reinforce a body of sand.

Split pattern. A pattern that is parted for convenience in molding.

Strike or strickle. A template or straightedge used for removing excess sand from a mold or core box.

Sprue. The opening into which the metal is first poured.

Stock cores. Standard cores of common diameters which are kept "in stock" for general use.

Stopping off. Closing off a part of the mold that is not wanted.

Stripping plate. A plate, formed to the contour of the pattern, which holds the sand in place while the pattern is drawn through the plate.

Sweep or skree. A board shaped to a required profile. It is used to remove excess material from a mold or core.

Sweep work. Forming molds or core by the use of jigs or templates instead of patterns.

Tucking. Pressing the sand in place with the hands.

Vent. Small opening in mold to facilitate escape of air and gases.

Vibrator. A mechanical device used to loosen pattern from mold.

APPENDIX II

GATING AND RISERING NOMENCLATURE

The Gating and Risering Committees of the American Foundrymen's Society have done a great deal toward standardizing the nomenclature connected with the feeding of castings. Hence the definitions evolved by these groups serve as a useful reference for this purpose.

The basic and very common gating system consists of the following elements. The metal is poured through the downsprue, enters the runner, and passes through the ingates into the mold cavity (Fig. A–1). That part of the gating system which most restricts or regulates the rate of pouring is the primary choke, more often called simply the choke. At the top of the downsprue may be a pouring cup or pouring basin to minimize splash and turbulence and ensure that only clean metal enters the downsprue. To further prevent the entry of dirt or slag into the downsprue, the pouring basin may contain a skim core, a strainer, a delay screen, or a sprue plug. To prevent erosion of the gating system when a large amount of metal is to be poured, a splash core may be placed in the bottom of the pouring basin, at the bottom of the downsprue, or wherever the flowing metal impinges with more than normal force.

Castings of heavy sections or of high-shrinkage alloys commonly require a riser or reservoir where metal stays liquid while the casting is freezing. The riser thus provides the feed metal which flows from the riser to the casting to make up for the shrink which takes place in the casting metal as it changes from liquid to

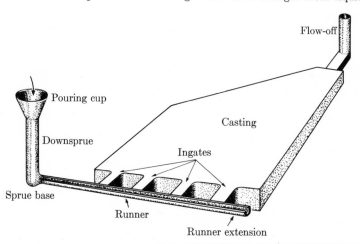

Fig. A–1. Finger-gated casting with flow-off. (Courtesy of A.F.S. Gating and Risering Committee)

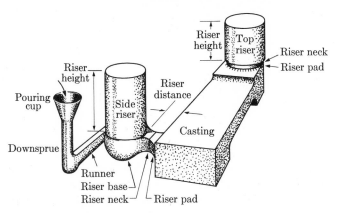

FIG. A–2. Riser-gated casting with side riser and blind riser. (Courtesy of A.F.S. Gating and Risering Committee)

solid. Depending on the location, the riser is described as a top riser or side riser and may be either an open riser or a blind riser. Since risers are designed to stay liquid while the casting solidifies, riser height and riser neck are important dimensions as are those of the body of the riser itself. Riser distance and the shape of the riser base are additional important details that pertain only to side risers.

Gates and risers are often designed to take advantage of the principle of controlled directional solidification which requires that freezing start farthest from the riser and proceed toward the riser. To accomplish this, castings are riser gated; the metal enters the riser through a downsprue and runner and heats both the riser base and riser neck while flowing into the mold cavity (Fig. A–2). A good general measure of whether gates and risers have the proper dimensions may be obtained by calculating casting yield.

Gating system (gates). In everyday terms, the arrangement of "plumbing" which conducts metal into the mold cavity.

Downsprue (sprue, downgate). The first channel, usually vertical, which the metal enters; so called because it conducts metal down into the mold.

Sprue base (button). An enlargement or rounded section at the bottom of the downsprue, used to help streamline the flow of metal into the runner.

Runner (crossgate). The second channel, usually horizontal, through which the metal flows toward, or is distributed around, the mold cavity.

Ingate (gate). The third channel through which the metal leaves the runner to enter either the mold cavity or riser adjacent to the cavity.

Runner extension. That part of a runner which extends beyond the farthest ingate as a blind end. It acts as a dirt trap since the first rush of metal along the runner will pick up any loose particles of sand or dirt and carry them into the extension and not into the mold cavity.

Mold cavity. The hole which when filled with metal becomes the casting. Gates and risers are not considered part of the mold cavity.

Primary choke or choke. That part of the gating system which most restricts or regulates the flow of metal into the mold cavity.

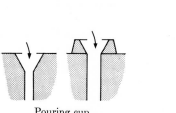

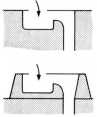

Pouring cup   Pouring basin

Pouring cup. The flared section of the top of the downsprue. It may be shaped by hand in the cope, or may be a shaped part of the stick used to form the down-sprue, or may be a baked core cup placed on top of the cope over the downsprue.

Pouring basin. A basin or trough into which the metal is poured and from which it enters the downsprue. It may be cut into the cope or may be a baked core basin set on top of the mold, or it may be a trough rammed up in a wood or metal frame (when it is desired that the metal enter two or more downsprues at the same time).

Skim core (skimmer). A flat core or tile placed to skim a flowing stream of metal. Commonly used in pouring basins, it holds back slag and dirt while clean metal passes underneath to the downsprue. Skim cores are also used elsewhere in gating systems.

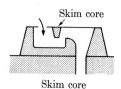

Skim core

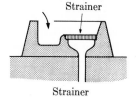

Strainer

Skim core   Strainer

Strainer. A core (strainer core), or tile (strainer tile), or ceramic-coated screen with many small holes. Used in a pouring basin, the strainer restricts flow of metal so that the basin fills quickly and only clean metal enters the downsprue. Strainers are also used at the bottom of the downsprue and elsewhere in the gating system to restrict the flow of metal.

Delay screen (skim gate, skim strainer). A small piece of perforated light-gauge tinned sheet steel frequently placed in the pouring basin at the top of the downsprue. The screen delays the flow of metal long enough to allow the basin to fill before it melts. Thus only clean metal from the bottom of the basin is permitted to enter the downsprue. Delay screens are also used elsewhere in the gating system.

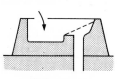

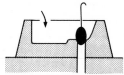

Delay screen   Sprue plug   Splash core

Sprue plug (stopper). A stopper placed at the entrance of the downsprue. Usually of cast iron or graphite, it ensures pressurized flow of clean metal into the downsprue since it is pulled only after pouring has started, when the basin is full of metal.

Splash core. A core or tile placed to prevent erosion of the mold at places where metal impinges with more than normal force. Splash cores are commonly used at the bottom of large rammed pouring basins, at the bottom of long downsprues, or at the ingates of large molds.

Riser (head, feed head). A reservoir connected to the casting for the purpose of feeding liquid metal to the casting during solidification, to offset the shrinkage which takes place while the casting is solidifying.

Shrink (draw, pull). The difference in volume between liquid metal and solid metal or the hole left in a casting because of that difference. Castings of high-shrinkage alloys require that extra metal flow into the casting during solidification; otherwise there will be a shrinkage cavity in place of the last metal to solidify.

Feed metal. The volume of liquid metal that passes from the risers to the casting to make up for the shrink.

Top riser (top head). A riser attached to the top surface of a casting.

Side riser (side head). A riser attached to the side of a casting.

Open riser (open head). A riser which cuts through the cope surface of a mold.

Blind riser (blind head, blind bob, shrink bob). A hidden riser, not visible at the top surface of the cope. Blind risers generally are more efficient and less expensive than open risers, but can be used only under certain conditions.

Riser height. The distance from the top of the riser when liquid to the top of the riser neck. The height of the riser when solid may be several inches less because of loss of feed metal to the casting.

Riser neck. The connecting passage between the riser and casting. Usually only the height and width, or diameter, of the riser neck are reported, although the shape may be equally important.

Riser pad (riser contact). An enlargement of the riser neck where it joins the casting. The purpose of the pad is to prevent the riser from "breaking in" when it is struck or cut from the casting.

Riser distance. The length of the riser neck. The term is applied to side risers only.

Riser base (drag bob). The shape of the bottom of a side riser.

Directional solidification (controlled freezing). A basic rule for the intelligent placing of risers. It requires that solidification start at the point farthest from the riser and proceed progressively toward the riser. With directional solidification, the riser can provide feed metal for each part of the casting as required, and the casting will be sound.

Riser gating. Practice of running metal for the casting through the riser to help directional solidification.

Casting yield. Weight of cleaned rough casting divided by total weight poured, expressed in per cent.

Finger gate (branch gate). A gating system with two or more ingates from the same runner. It distributes metal to several parts of the mold cavity.

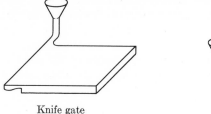
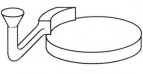

Knife gate   Lap gate

Step gate. A gating system with two or more ingates at different elevations from a vertical runner or downsprue.

Knife gate. A slit opening of thickness $\frac{3}{16}$ to $\frac{1}{4}$ in. and any length through which metal enters the mold cavity. Knife gates are usually poured on a downhill tilt to permit progressive filling of the mold cavity. Knife gates are easy to remove.

Lap gate. A slit opening $\frac{1}{16}$ to $\frac{3}{16}$ in. thick, and usually less than 3 in. in length which is formed by overlapping the runner with an edge of the mold cavity. Lap gates are easy to remove.

Horseshoe gate. A gating system in which a runner and two ingates are combined in the shape of a horseshoe. This gating system is easily hand-cut and removed from the casting.

Shower gate (pencil gate, pop gate). A gating system through which metal showers into the mold cavity from a group of small gates at the top.

Bottom gate. Any gating system through which metal enters the mold cavity at the bottom.

Horn gate. A curved bottom gate through which metal flows down under and then up into the mold cavity. A *reverse horn gate* is attached by the large end to the casting to reduce turbulence in the metal.

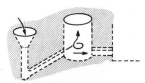

Horseshoe gate   Whirl gate

Whirl gate. A gating system in which the metal enters a circular reservoir at a tangent, and so whirls around, leaving dirt and slag behind before passing into the mold cavity.

Skim bob. A small upward bulge in the ingate an inch or two from the casting which acts as a dirt trap.

Neck down (wafer core, knock-off, Washburn, or Cameron core). A thin core or tile used to restrict the riser neck, making it easier to break or cut off the riser from the casting.

Vent (whistler). A small channel from the top of the mold cavity designed to let air and mold gases escape as the metal fills the mold.

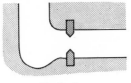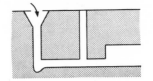

Neck down   Relief sprue

Flow-off (pop-off, strain relief). A large vent, usually located at the high point of the mold cavity. In addition to letting air and mold gases escape as metal fills the mold cavity, the flow-off fills with metal and acts to relieve the surge of pressure near the end of pouring.

Relief sprue. A vertical channel, the approximate size of the downsprue, connected to the runner to relieve pressure surge during pouring. It functions like a standpipe in a plumbing system.

Set gates. Gating systems formed by patterns, in contrast to gates cut in the sand by hand.

Pencil core. A core projecting to the center of a blind riser to admit atmospheric pressure for the purpose of forcing out feed metal.

INDEX

ABCDE69876543